T0276819

# Robotics: Application, Control Strategies and Coding

## Volume II

# Robotics: Application, Control Strategies and Coding Volume II

Edited by **Jared Kroff**

New Jersey

Published by Clanrye International,
55 Van Reypen Street,
Jersey City, NJ 07306, USA
www.clanryeinternational.com

Robotics: Application, Control Strategies and Coding
Volume II
Edited by Jared Kroff

International Standard Book Number: 978-1-63240-455-8 (Hardback)

Printed in the United States of America.

# Contents

**Permissions**

**List of Contributors**

# Preface

The purpose of the book is to provide a glimpse into the dynamics and to present opinions and studies of some of the scientists engaged in the development of new ideas in the field from very different standpoints. This book will prove useful to students and researchers owing to its high content quality.

This book combines recent researches in robot sensors and algorithms. It has been organized under two broad sections namely, "Vision and Sensors" and "Programming and Algorithms" focusing on ultrasonic-sensors, programming of intelligent robots and systems adaptations. The advancements of these novel techniques have given a strong stimulus in the field of robotics.

At the end, I would like to appreciate all the efforts made by the authors in completing their chapters professionally. I express my deepest gratitude to all of them for contributing to this book by sharing their valuable works. A special thanks to my family and friends for their constant support in this journey.

**Editor**

# Part 1

# Vision and Sensors

# Design and Construction of an Ultrasonic Sensor for the Material Identification in Robotic Agents

Juan José González España, Jovani Alberto Jiménez Builes
and Jaime Alberto Guzmán Luna
*Research group in Artificial Intelligence for Education, Universidad Nacional de Colombia*
*Colombia*

## 1. Introduction

Quality assurance in all industrial fields depends on having a suitable and robust inspection method which let to know the integrity or characteristics of the inspected sample. The testing can be destructive or not destructive. The former is more expensive and it does not guarantee to know the condition of the inspected sample. Though, it is used in sectors where the latter cannot be used due to the technical limitations. The second one is an ongoing research field where cheaper, faster and more reliable methods are searched to assess the quality of specific specimens.

Once a suitable reliability and quality are achieved by using a specific method, the next step is to reduce the duration and cost of the associated process. If it is paid attention to the fact that most of these processes require of a human operator, who is skilled in the labour of interpreting the data, it can be said that those previously mentioned factors (process duration and cost) can be reduced by improving the automation and autonomy of the associated processes. If a robotic system is used this can be achieved. However, most of the algorithms associated to those processes are computationally expensive and therefore the robots should have a high computational capacity which implies a platform of big size, reduced mobility, limited accessibility and/or considerable cost. Those constraints are decisive for some specific applications.

One important factor which should be considered to develop a low cost, small size and mobile robotic system is the design of the software and hardware of the sensor. If it is designed with a depth analysis of the context of the specific application it can be obtained a considerable reduction on the requirements and complexity of the robotic system.

The appropriated design of the hardware and software of the sensor depends on the proper selection of the signal pattern which is going to be used to characterize the condition of the inspected sample. For ultrasonic waves the patterns are changes on amplitude, frequency or phase on the received echo because of the properties of a specific target. Some of those properties are attenuation, acoustic impedance and speed of sound, among others.

Among the many applications of ultrasound, one of them which is important for the aerospace, automotive, food and oil industries, among others, is the material identification. Depict the fact that there are many ultrasonic sensors which let to identify the material of the sample being inspected, most of them are not appropriated to be implemented in a robot. The high computational capacity demanded by the algorithm of the sensor is the main constraint for the implementation of the ultrasonic identification of materials in a robot. In this book chapter is addressed this problem and it is solved by the Peniel method. Based on this method is designed and constructed a novel sensor which is implemented in two robots of the kit TEAC$^2$H-RI.

This book chapter is organized as follows: Section Two briefly reviews some approaches in Material Identification with ultrasonic sensors. Section Three illustrates the novel Peniel method and its associated hardware and software. Section Four describes the results. Finally, section five presents the conclusions and future work.

## 2. Material identification with ultrasonic sensors

The material identification using ultrasonic techniques is very important for a plethora of sectors of the industry, research, security, health, among others. In contrast with other methods for the identification of materials, those based on ultrasonic signals not only are cheaper but also some of them let to identify objects which are occluded by others.

Following is the general procedure to achieve this:

1. It is used a specific configuration for the receiver and transmitter ultrasonic transducers. Some of them are: Pulse-Echo, Through-Transmission and Pitch-Catch configuration [NASA, 2007]
2. It is sent an ultrasonic wave to the inspected sample
3. Once the wave has interacted with the object, it returns as an echo to its source and there it is captured by means of an ultrasonic transducer (receiver).
4. Then, the electrical signal goes to a signal conditioning system, which is formed by amplifiers, filters, digital to analog converters (DAC), and so on.
5. Finally, there is the processing module, which is in charge to perform the most important stage of the system, i.e., the signal processing. In this case, it can be used microcontrollers, microprocessors, FPGAs, computers, among other devices. In this processing module the algorithm is implemented, in order to process the respective signal patterns and indicates the material under inspection.

In the next lines is illustrated some methods proposed in the state of the art for the material identification and is mentioned their strengths and more important limitations.

In [Thomas et al, 1991] is developed an ultrasonic perception system for robots for the task of identification of materials. This system is based on the neural network Multi Layer Perceptron (MLP), which was trained with the characteristics of the frequency spectrum and the time domain of the ultrasonic echo. Its most relevant advantage is that the sample was inspected from different positions; additionally the ultrasonic transducer did not require making contact with the inspected sample. Its more remarkable limitations are that it identifies the material in a stochastic manner, i.e. sometimes the material is correctly identified and other not, and the issue that the sensor was never implemented in a robotic system.

In [Gunarathne et al, 2002] is proposed an implementation to identify the material of nonplanar cylindrical surface profiles (e.g. petroleum pipelines) with a prediction accuracy of the result. The method is mainly based on the feature extraction by curve fitting. The curve fitting consists in a computationally generated artificial signal which fits the shape of the experimental signal. The optimal parameters chosen for the artificial signal (AS) are those which make the AS to fit the experimental signal with a low error; those are used for the task of material identification.

Furthermore, for this case is used a database which contains the typical values of the parameters correspondent to specific materials. Its more relevant strengths are the identification of the materials of objects of non-planar cylindrical surface profiles and the accuracy of the results. On the other hand, its more significant limitations are the expensive computational cost of its processing algorithm, the absence of autonomy and the restrictions to the position of the transducer.

In [Ohtanil et al, 2006], is implemented an array of ultrasonic sensors and a MLP neural network for the materials identification. Some experiments were performed on copper, aluminium, pasteboard and acrylic, at different angles with respect to the x and z axes. The maximum distance between the sensor array and the target was 30cm. The main strength of the system is that it works in air over a wide range of distances between the target and the sensor system. Additionally, the sensor model is automatically adjusted based on this distance. The results are quite accurate. Its more significant limitation is the big size of the system (only appropriated for a huge robot), the computing requirements and the dependence on the position of the target to achieve the identification of the material.

In [Zhao et al, 2003] is used the ultrasonic identification of materials for the quality assurance in the production of canned products. The configuration of the ultrasonic system is pulse-echo. Water is the coupling medium between the transducer and the target. By means of this set up, it can be detected any shift from a specific value of the acoustic impedance, which is the result of foreign bodies within the bottle. Its main strengths are the accuracy of the system and the high sensibility. Its more relevant weaknesses are the absence of mobility and autonomy.

In [Pallav et al, 2009] is illustrated a system which has the same purpose of the previous mentioned system, i.e. also by means of the material identification performs quality control in canned food. The system uses an ultrasonic sensor to identify the acoustic impedance shift within the canned food which is related to foreign bodies within the can. The main difference in comparison to the previous mentioned system [Zhao et al, 2003] is the fact that the Through-Transmission configuration is used, and that the couplant is not water but air. Its more relevant strengths are the successful detection of foreign bodies and the mobility achieve in the X and Y axes. Its more important limitations are the high requirements of voltage, the dependence of the object's position and the narrow range of mobility.

In [Stepanić et al, 2003] is used the identification of materials to differentiate between common buried objects in the soil and antipersonnel landmines. For this case is used a tip with an ultrasonic transducer in its border and this latter is acoustically coupled with the target. The procedure to operate the system is: 1. Select a zone where is suspected the landmine is. 2. Introduced the tip in this zone in order to touch the suspecting object 3. Take some measurements. Based on this measurements the material can be identified and as a consequence it can be detected the presence or absence of landmines. Its more important limitations are the lack of autonomy, because it requires a human operator, and the dangerous fact which involves making contact with the landmine.

From the previous mentioned approaches the most significant limitation which forbids the automation in small robotic agents is the expensive computational cost of the processing algorithms. This problem was addressed in a master thesis and the results are exposed in this article.

## 3. Peniel method

In the following lines is exposed the Peniel Method, which is a method of low computational cost and therefore its associated algorithm and circuit can be implemented in a small robot. It is important to highlight that only one microcontroller is used to implement this method.

### 3.1 Mathematical model

In [Gunarathne et al, 1997, 1998, 2002; Gonzalez & Jiménez, 2010b] is exposed that the reverberations within a plate follow the next expression.

$$A(t) = Ae^{-B(t-C)} + D \tag{1}$$

where A is related to the first echo amplitude, B to the rate of decay of reverberations, C to the timing of the first echo from the start of the scan and D to the signal-to-noise ratio.

In [Gunarathne et al, 2002] is mentioned that knowing the decay coefficient (B) the material can be identified. Moreover, in [Allin, 2002] is illustrated that there are several methods that are based on the decay coefficient to identify the material. This variable is important because it only depends on the attenuation and acoustic impedance of the material, if the thickness is kept constant. In most of the solids, the attenuation is low and its effect is negligible in comparison with the acoustic impedance. Based on this the material can be identified if the acoustic impedance is known. In the following lines we present a novel method developed to estimate indirectly the decay coefficient B and therefore it can be used to identify the material of solid plates.

Without losing generality, in (1) we assume that D=0 and C=0. If it is sought the time interval during $A(t)$ is greater than or equal to a value $w$, then the equation which express this is:

$$A(t) \geq w \tag{2}$$

Replacing (1) in (2) and taking into account the considerations that were made with respect C and D, we obtained:

$$\leftrightarrow Ae^{-B \cdot t} \geq w \tag{3}$$

After some algebra we obtained:

$$\leftrightarrow t \leq \frac{\ln\left(\dfrac{A}{w}\right)}{B} \tag{4}$$

It means that in the time interval $\left[0, \dfrac{\ln\left(\dfrac{A}{w}\right)}{B}\right]$ the exponential function is greater than or

equal to $w$. The duration of this time interval is defined as $t_{di,}$ and then for the previous case it can be said that:

$$t_{di} \leq \frac{\ln\left(\dfrac{A}{w}\right)}{B} \tag{5}$$

The left side term in the expression (3) is an artificial signal which fits the received ultrasonic echo (see page 3). If the received ultrasonic echo is amplified by a Gain G, as will the exponential function (3). Thus it is obtained from (3):

$$G \cdot A e^{-B \cdot t} \geq w \tag{6}$$

It leads to expressions similar to those obtained in (4) and (5):

$$t \leq \frac{\ln\left(\dfrac{G \cdot A}{w}\right)}{B} \tag{7}$$

$$t_{di} = \frac{\ln\left(\dfrac{G \cdot A}{w}\right)}{B} \tag{8}$$

Therefore, for two signals with similar A values, the $t_{di}$ duration is inversely proportional to the decay coefficient B. It can be seen in figure 1 for the case when A = 1, $w$ = 0.5, and B = 1000, 2000, 3000 and 4000. Additionally, it is shown the behavior with G of the difference between $t_{di3}$ and $t_{di4}$. This let us to conclude that $t_{di}$ can be used to characterize a material if G, A and $w$ are known.

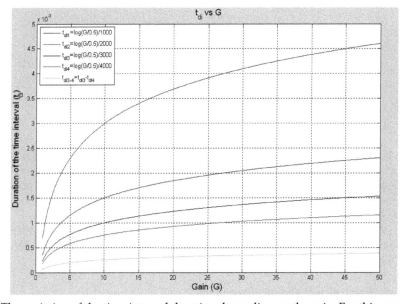

Fig. 1. The variation of the time interval duration depending on the gain. For this case, the time interval is the time during an exponential signal is above 0.5 ($w$)

Now, if it is used two different gain values (G1 and G2) for the same signal, the equation (8) yields:

$$t_{di1} = \frac{\ln\left(\dfrac{G_1 \cdot A}{w}\right)}{B}$$

(9)

$$t_{di2} = \frac{\ln\left(\dfrac{G_2 \cdot A}{w}\right)}{B}$$

(10)

If $G_1 < G_2$ and (9) is subtracted from (10), it is obtained:

$$t_{di2} - t_{di1} = \frac{\ln\left(\dfrac{G_2 \cdot A}{w}\right)}{B} - \frac{\ln\left(\dfrac{G_1 \cdot A}{w}\right)}{B}$$

(11)

$$t_{di2} - t_{di1} = \frac{\ln\left(\dfrac{G_2 \cdot A}{w}\right) - \ln\left(\dfrac{G_1 \cdot A}{w}\right)}{B}$$

(12)

$$t_{di2} - t_{di1} = \frac{\ln\left(\dfrac{\dfrac{G_2 \cdot A}{w}}{\dfrac{G_1 \cdot A}{w}}\right)}{B}$$

(13)

$$t_{di2} - t_{di1} = \Delta t_1 = \frac{\ln\left(\dfrac{G_2}{G_1}\right)}{B}$$

(14)

$$\Delta t_1 = \frac{\ln(G_r)}{B}$$

(15)

where $G_r = \dfrac{G_2}{G_1}$

As it can be seen (14) does not depend on A and therefore is only necessary to know $G_2$ and $G_1$ to find B from $\Delta t_1$. Moreover, if there are two signals from different materials, even though $G_2$ and $G_1$ are not known, the material can be identified from the difference between the correspondent $\Delta t$s. In figure 2 is shown the behavior of $\Delta t$ as a function of B for different values of $G_r$.

As it can be seen in the figure 2 the duration increment of the time interval ($\Delta t$) depends on B if $G_r$ is kept constant. Also, it can be recognized that if B is kept constant, high values of $G_r$ result in high increases on the duration of the time interval ($\Delta t$).

### 3.2 The electronic circuit and the algorithm

One of the conditions to identify the material using the expression (8) of the Peniel Method, is that those signals compared in terms of the behavior of the time interval durations should have a similar value A (see equation 1). To meet this requirement it has been used the circuit in the figure 3. This circuit is called **clustering circuit**. Despite the fact that by means of (14) the material can be identified without knowing A, the clustering circuit is used to make more robust the method.

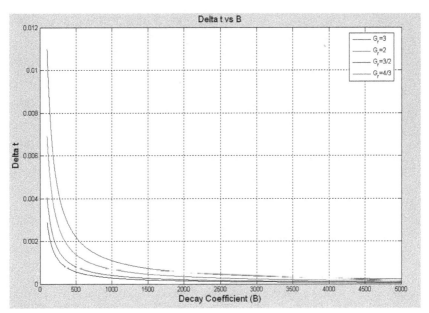

Fig. 2. Behavior of the duration increment of the time interval, ($\Delta t$), in function of the decay coefficient for values of $G_1$=10, 20, 30 and $G_2$= 20, 30, 40

### 3.2.1 Clustering algorithm and circuit

This circuit is formed by four amplifiers, four comparators and a microcontroller. Each amplifier has different gains. The gain grows in the direction of AMP4, i.e., the smaller and higher gains belong to the AMP1 and AMP4 amplifiers, respectively. Each amplifier represents a group, in this manner AMP3 represents the group 3 and group 2 is represented by AMP2

The output of the amplifiers goes to the respective comparators. The threshold ($w$) is the same for the last three comparators and a little bit higher for the first comparator (AMP1). The echo signal belongs to the group which has the amplifier with lower gain but with its comparator activated. In this manner a signal which activates the comparator three only belongs to the group three if the respective comparators of AMP1 and AMP2 are not activated.

The output of each comparator goes to the microcontroller and once this detects low levels on any comparator, it identifies what is the amplifier with lower gain which has its comparator activated and therefore it identifies the group the signal belongs to.

Consequently, those signals with similar peak amplitude, A (see equation 1), will be in the same group. In this manner is met the requirement that the A value should be similar between signals which are compared in terms of the duration of the time interval.

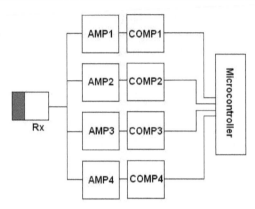

Fig. 3. Clustering circuit to assure that the A value is similar in signals which are compared in terms of duration of the time interval. The abbreviation Amp refers to the amplifier and Comp to the comparator

### 3.2.2 The Peak voltage detection circuit and the algorithm

In addition to the previous mentioned circuit also the system has a circuit that allows knowing the peak value of the received echo. This peak value is an estimation of A, which is also a parameter of the expression (1) and therefore can be used to identify the material. The associated circuit is shown in Figure 4. The AMP1, 2, 3 and 4 are the same as in the circuit of Figure 3. The ENV1, 2, 3 and 4 are envelope detectors, whose function is conditioning the signal before is sent to the microcontroller. The CD4052 is a multiplexer. The exit of each comparator goes to a specific input of the CD4052, and the output of this latter in the pin 3 goes to the microcontroller.

As mentioned above, the clustering circuit identifies which group the signal belongs to; this information is used by the microcontroller to select which of the four envelope detectors should be used to find the peak amplitude. This selection is done by means of the CD4052. Once the microcontroller has chosen the right envelope detector and captured the correspondent signal, it proceeds to search the peak value by successive analog to digital conversions and comparison.

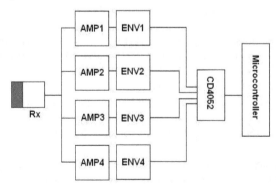

Fig. 4. Electronic circuit for calculating an estimated value of A. AMP1, AMP2, AMP3 and AMP4 are the same as in Figure 3. ENV means envelope detector

### 3.2.3 The Circuit and the algorithm for the identification of the changes in the durations

As it was mentioned in section 3.1 the material can be identified by finding the difference between the two durations which are produced for the same signal when it is amplified by two different gain values, i.e. finding $\Delta t$.

In order to do this, the circuit of Figure 5 was implemented. In this case AMP1-4 correspond to the same amplifiers mentioned in Figure 3. Amp1.1, Amp2.1, Amp3.1 and Amp4.1 are additional amplifiers. COMP5, COMP6, COMP7 and COMP8 are comparators which even with an amplitude modulated sinusoidal signal, such as the ultrasonic echo, remains activated as long as the envelope of the signal is above the threshold of 2.5V. While the output of this comparator is quite stable, still it has some noisy fluctuations which are necessary to be removed and this is the reason to use the comparators COMP5.1, COMP6.1, COMP7.1 and COMP8.1. The output of these comparators is connected to a specific input in the CD4052, and the output in the pin 13 of this multiplexer goes to the microcontroller.

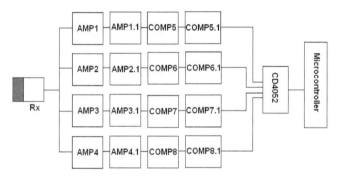

Fig. 5. The electronic circuit for calculating the durations associated to different gain values

## 4. Results and discussion

In order to prove the effectiveness of the Peniel Method, it was performed some experiments. The characteristics of the experiments are mentioned below:

The transducer used for the measurements was the AT120 (120KHz) of Airmar Technology, which is a low cost transducer commonly used for obstacle avoidance or level measurement but not for NDT. This transducer was used both for transmission and reception in the Through-Transmission configuration which was used to inspect the samples.

The transmitter circuit is the T1 *development board* of Airmar Technology (Airmar Technology, 2011). With this transmitter is sent a 130KHz toneburst which has a 40µs duration and a pulse rate of 10Hz.

The chosen samples for the inspection were Acrylic (16mm thickness), Aluminium (16mm thickness) and Glass (14mm thickness). Oil was the couplant used between the transducers and the inspected sample.

The receiver circuit is the result of integrating the three circuits mentioned in the section 3.2. The circuit is fed by four 9volts batteries. Two of them are for the positive source and the other two are for the negative. The data were processed by the main board of the mom-robot of the kit TEAC²H-RI (González J.J. *et al*, 2010a).which was developed by the authors. In the main board, the microcontroller MC68HC908JK8 is the device in charge to assign tasks or process the data of the sensors. The six outputs from the receiver circuit go to the main board and two outputs of the main board go to the receiver circuit. These two outputs control the CD4052.

The microcontroller sends the duration and the peak voltage of the signals to the PC. There, the signal is processed by Matlab and the results are plotted.

The whole set up of the experiments can be seen in figure 6.

In order to assure the repeatability of the results the following procedure was followed:

**Procedure 1**

1.  Clean the inspected sample and the transducers face
2.  Spread oil over the surface of both transducer faces
3.  Push the transducers against the inspected sample
4.  Take ten different measurements
5.  Repeat this procedure for five times.

The ten different measurements refer to send in different moments a toneburst to the inspected sample and capture the received echo. This is done ten times.

For the different materials used in this experiment this procedure was followed and the results are in figure 7.

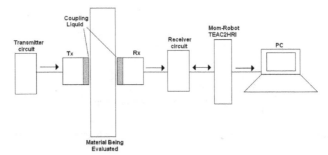

Fig. 6. Diagram of the Experimental set up used in procedure 1. The couplant is oil. Tx and Rx refer to the Transmitter and Receiver, respectively

In figure 7 Delta $t_1$ refers to $t_{di2}$ - $t_{di1}$, Delta $t_2$ refers to $t_{di3}$ - $t_{di2}$, and so on. $t_{di1}$ refers to the duration of the signal at the output of the comparator which corresponds to the group signal and the other $t_{di,}$ belong to the comparators of the groups with amplifiers with higher gain than the amplifier of the group of the signal, e.g. if the signal belongs to group 2 the $t_{di1}$ refers to the duration of Comp6.1, and $t_{di2}$ and $t_{di3}$ belong to Comp7.1 and Comp8.1, respectively. As it can be seen, for the higher order groups the number of $t_{di}$ will be fewer than for the groups of lower order.

As it can be seen from figures 7a-7b, the $\Delta t$ is not constant for any material. However, it can be said that this value is always within a defined interval. In [Gunarathne et al, 2002] is mentioned that the decay coefficient B does not have a fixed value from measurement to measurement but it takes different values within an interval. Because of this is defined a Gaussian Probability Density Function for the B value of each material, then the value can be any of those values within the PDF. This fact, confirms that the results obtained here with $\Delta t$ are correct because they are congruent with the theory. Though, for the current case it was not necessary to create a PDF but only intervals where the correspondent $\Delta t$ belongs. This fact simplifies the material identification task. Following, the intervals are chosen.

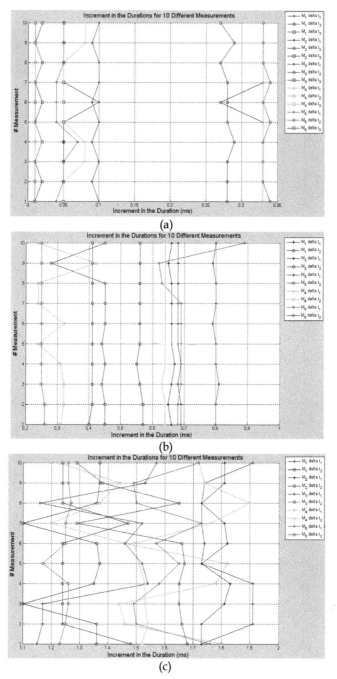

Fig. 7. Duration increment of the time interval for the results obtained using the procedure 1. The materials inspected were a) Acrylic b) Glass and c) Aluminum

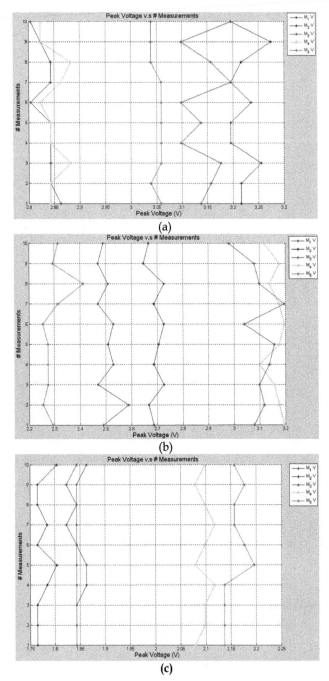

Fig. 8. Peak voltage for the results obtained using the procedure 1. The materials inspected were a) Acrylic b) Glass and c) Aluminum

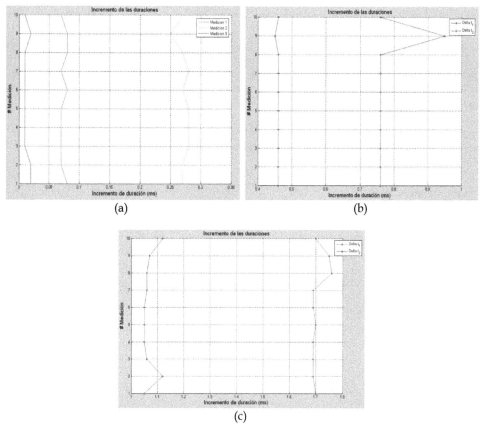

Fig. 9. Duration increment of the time interval for the results obtained using the procedure 2. The materials inspected were a) Acrylic b) Glass and c) Aluminum

From figure 7a it can be seen that the $\Delta t_1$ is always within the range (0.025ms, 0.15ms), while $\Delta t_2$ and $\Delta t_3$ are in the ranges (0.16ms, 0.4ms) and (0, 0.1ms), respectively, then these intervals are chosen to identify the acrylic sample.

From figure 7b it can be seen that the $\Delta t_1$ and $\Delta t_2$ are always between (0.6ms, 1ms) and (0.2ms, 0.65ms), respectively. Then, those are the intervals for the glass sample.

From figure 7c it can be seen that the $\Delta t_1$ and $\Delta t_2$ are always between (1ms, 2ms) and (0.8ms, 1.9ms), respectively. Then, those are the intervals for the aluminium sample.

In figure 8 is shown the peak voltages for the correspondent measurements in figure 7. It can be seen from this figure that the peak voltage also varies in a wide range from measurement to measurement. Then, if this parameter is going to be implemented to identify the material also have to be defined a voltage interval for each material. This parameter can be used to make more robust the measurements but for this moment only the $\Delta t$ variable will be used for the identification of the materials.

It is important to mention that both, the peak voltage and $\Delta t$ change, are within an interval from measurement to measurement because those variables are very dependent on the characteristics of the couplant used between the transducer face and the inspected sample.

From the previous mentioned results the algorithm was modified and the intervals were used to identify the material. With this new algorithm a procedure (procedure 2) similar to procedure 1 was repeated for all the materials. The only difference in this new procedure is that the average of the ten measurements is taken and this result is compared with the different intervals. In figure 9 can be seen one of the five repetitions of the new procedure.
In table 1 can be seen the performance of the developed sensor for the material identification task.

| Material | Trials | Correct Identification | Wrong Identification | % Accuracy |
|---|---|---|---|---|
| Acrylic | 5 | 5 | 0 | 100% |
| Glass | 5 | 5 | 0 | 100% |
| Aluminum | 5 | 5 | 0 | 100% |

Table 1. Accuracy of the developed sensor to identify the material

As it can be seen in these experiments the accuracy of the identification is 100% for all the materials. This fact is very important because with a quite easy method and a low cost computational algorithm was performed a task which requires of more complex methods and algorithms of more expensive computational cost.

### 4.1 The robotic kit TEAC$^2$H-RI
In the previous sections, the signal was processed using the main board of the mom-robot of the kit TEAC2H-RI (González J.J. et al, 2010a) joint with the software Matlab. Then, the Peniel Method was partially implemented in a robotic system.
In future work the sensor will be fully implemented in a robotic system formed by two robots of the kit TEAC2H-RI. These robots are the mom-robot and the son-robot. Both of them can be seen in figure 10

Fig. 10. Son-robot and mom-robot exploring the environment (González J.J. et al, 2010a)

## 5. Conclusion and future work

In this book chapter was proposed for the first time in the literature the shift in $\Delta t$ as a variable which can be used to identify the material. The method based on this variable, Peniel Method, is much easier and much less computationally expensive than other methods in the literature. This fact is the most important approach of this research because it lets to implement the ultrasonic material identification sensor in two robots. This last result was not found in the literature reviewed.

The cost of the ultrasonic AT120 transducer is only $35US which is very low in comparison with other options (>$400US) in the market which are used for the identification of materials. Then, it can be concluded that low cost ultrasonic transducers can be used instead of more expensive transducers if the proper method is used. This is another important strength of the method because it can be used in industrial processes to automate the material identification task at a low cost.

Another important conclusion of this work is that if the appropriate methodology is followed to design a sensor, it could be save considerable economical investments in the transducer, the processing system and the other elements necessary to design the sensor. In this case the methodology followed to develop the sensor consists of seven steps: 1) to study in depth the model or physical-mathematical models proposed in the literature 2) to perform experiments to confirm the theory and identify new signal patterns which let to develop new models 3) to redefine or create a new physical-mathematical model which must be simpler than the others and fits to the specific characteristics of the application, which in this case is two small robots to identify materials 4) This new model is the basis of the new method, but again it is necessary to also consider the characteristics of the specific application and all the constraints in order to design the electronic circuit and the algorithm of the method 5) to build the sensor with the proper devices in terms of cost, size and functionality 6) to design the appropriate experiments for the verification of the method 7) to validate the performance of the sensor using the proposed experiments and debug both the algorithm and circuit for optimum performance.

As it can be seen not only is necessary to design or redefined the software or hardware but also the mathematical model in order to obtain the sensor which fulfils the requirements of the application at the lowest cost.

The next step of this work is to implement completely the sensor in the mom-robot and son-robot of the kit TEAC$^2$H-RI without depending on the processing of the PC to identify the material.

Also more materials will be evaluated with the sensor in order to provide to the robot a wider database of materials.

Also in future work, will be used the peak voltage as another factor to identify the materials or to differentiate between materials of very similar acoustic properties.

## 6. Acknowledgment

We would like to acknowledge to the Science, Technology and Innovation Management Department, COLCIENCIAS, for the economical support to the master student Juan José González España within the Youth Researchers and Innovators Program- 2009.

# 7. References

Allin M. J. (2002) Disbond *Detection in Adhesive Joints Using Low Frequency Ultrasound*. PhD Thesis. Imperial College of Science Technology and Medicine. University of London. 2002

Airmar Technology. (2011). http://www.airmartechnology.com/ Accesed 4th August 2011

González J.J.; Jiménez J.A.; Ovalle D.A. (2010) *TEAC2H-RI: Educational Robotic Platform for Improving Teaching-Learning Processes of Technology in Developing Countries*. Technological Developments in Networking, Education and Automation 2010, pp 71-76

González J., Jiménez J. *Algoritmo para la Identificación de Materiales en Agentes Robóticos (Spanish)*. Quinto Congreso Colombiano de Computación (5CCC) Cartagena-Colombia (2010)

Gunarathne, G. P. P. (1997) *Measurement and monitoring techniques for scale deposits in petroleum pipelines*. In: Proc. IEEE Instrum. Meas. Technol. Conf., Ottawa, ON, Canada, May 19–21, 1997, pp. 841–847. DOI: 10.1109/IMTC.1997.610200

Gunarathne G.P.P ; Zhou Q. ; Christidis K. (1998) *Ultrasonic feature extraction techniques for characterization and quantification of scales in petroleum pipelines*. In: Proceedings IEEE Ultrasonic Symposium, vol. 1; 1998, p. 859–64 DOI: 10.1109/ULTSYM.1998.762279

Gunarathne G.P.P ; Christidis K. (2002) *Material Characterization in situ Using ultrasound measurements*. In: Proceedings of the IEEE Transactions on Instrumentation and Measurement. Vol. 51, pp. 368-373. 2002 DOI: 10.1109/19.997839

NASA (2007) *Ultrasonic Testing of Aerospace Materials*. URL: klabs.org/DEI/References/design_guidelines/test_series/1422msfc.pdf Accessed on: August 2011.

Ohtanil K., Baba M. (2006) *An Identification Approach for Object Shapes and Materials Using an Ultrasonic Sensor Array*. Proceedings of the SICE-ICASE International Joint Conference, pp. 1676-1681. 2006

Pallav P., Hutchins D., Gan T. (2009) *Air-coupled ultrasonic evaluation of food materials*. Ultrasonics 49 (2009) 244-253

Stepanić, H. Wüstenberg, V. Krstelj, H. Mrasek (2003) *Contribution to classification of buried objects based on acoustic impedance matching*, Ultrasonics, 41(2), pp. 115-123, 2003

Thomas, S.M, Bull D.R, (1991) *Neural Processing of Airborne Sonar for Mobile Robot Applications*. Proceedings of the Second International Conference on Artificial Neural Networks. pp. 267-270. 1991.

Zhao B., Bashir O.A., Mittal G.S., (2003) *Detection of metal, glass, plastic pieces in bottled beverages using ultrasound*, Food Research International 36 (2003) 513–521.

# 2

# Robotics Arm Visual Servo: Estimation of Arm-Space Kinematics Relations with Epipolar Geometry

Ebrahim Mattar
*Intelligent Control & Robotics, Department of Electrical and Electronics Engineering,*
*University of Bahrain*
*Kingdom of Bahrain*

## 1. Introduction

### 1.1 Visual servoing for robotics applications

Numerous advances in robotics have been inspired by reliable concepts of biological systems. Necessity for improvements has been recognized due to lack of sensory capabilities in robotic systems which make them unable to cope with challenges such as anonymous and changing workspace, undefined location, calibration errors, and different alternating concepts. Visual servoing aims to control a robotics system through artificial vision, in a way as to manipulate an environment, in a similar way to humans actions. It has always been found that, it is not a straightforward task to combine "Visual Information" with a "Arm Dynamic" controllers. This is due to different natures of descriptions which defines "Physical Parameters" within an arm controller loop. Studies have also revealed an option of using a trainable system for learning some complicated kinematics relating object features to robotics arm joint space. To achieve visual tracking, visual servoing and control, for accurate manipulation objectives without losing it from a robotics system, it is essential to relate a number of an object's geometrical features (object space) into a robotics system joint space (arm joint space). An object visual data, an play important role in such sense. Most robotics visual servo systems rely on object "features Jacobian", in addition to its inverse Jacobian. Object visual features inverse Jacobian is not easily put together and computed, hence to use such relation in a visual loops. A neural system have been used to approximate such relations, hence avoiding computing object's feature inverse Jacobian, even at singular Jacobian postures. Within this chapter, we shall be discussing and presenting an integration approach that combines "Visual Feedback" sensory data with a "6-DOF robotics Arm Controller". Visual servo is considered as a methodology to control movements of a robotics system using certain visual information to achieve a task. Visionary data is acquired from a camera that is mounted directly on a robot manipulator or on a mobile robot, in which case, motion of the robot induces camera motion. Differently, the camera can be fixed, so that can observe the robot motion. In this sense, visual servo control relies on techniques from image processing, computer vision control theory, kinematics, dynamic and real time computing.

Robotics visual servoing has been recently introduced by robotics, AI and control communities. This is due to the significant number of advantages over blind robotic systems. Researchers have also demonstrated that, VISUAL SERVOING is an effective and a robust framework to control robotics systems while relying on visual information as feedback. An image-based scheme task is said to be completely performed if a desired image is acquired by a robotic system. Numerous advances in robotics have been inspired by the biological systems. In reference to Fig. (1), visual servoing aims to control a robotics system through an artificial vision in a way as to manipulate an environment, comparable to humans actions. Intelligence-based visual control has also been introduced by research community as a way to supply robotics system even with more cognitive capabilities. A number of research on the field of intelligent visual robotics arm control have been introduced. Visual servoing has been classified as using visual data within a control loop, enabling visual-motor (hand-eye) coordination. There have been different structures of visual servo systems. However, the main two classes are; Position-Based Visual Servo systems (**PBVS**), and the Image-Based Visual Servo systems (**IBVS**). In this chapter, we shall concentrate on the second class, which is the Image-based visual servo systems.

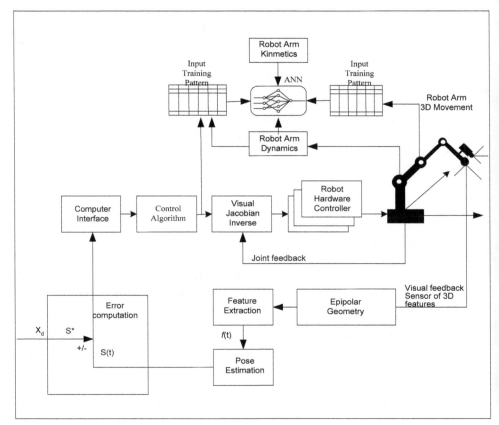

Fig. 1. Overall structure of an ANN based visual servoing

## 1.2 Literature surveys

EGT, Epipolar Geometry Toolbox, (Eleonora et al., 2004), was also built to grant MATLAB users with a broaden outline for a creation and visualization of multi-camera scenarios. Additionally, to be used for the manipulation of visual information, and the visual geometry. Functions provided, for both classes of vision sensing, the PINHOLE and PANORAMIC, include camera assignment and visualization, computation, and estimation of epipolar geometry entities. Visual servoing has been classified as using visual data within a control loop (Cisneros, 2004), thus enabling visual-motor (Hand-Eye) coordination.

Image Based Visual Servoing (**IBVS**) Using Takagi-Sugeno Fuzzy Neural Network Controller has been proposed by (Miao et. al, 2007). In their study, a T-S fuzzy neural controller based IBVS method was proposed. Eigenspace based image compression method is firstly explored which is chosen as the global feature transformation method. Inner structure, performance and training method of T-S neural network controller are discussed respectively. Besides that, the whole architecture of TS-FNNC was investigated.

An Image Based Visual Servoing using Takagi-Sugeno fuzzy neural network controller has been proposed by Miao, (Miao et. al, 2007). In this paper, a TAKAGI-SUGENO Fuzzy Neural Network Controller (TSFNNC) based Image Based Visual Servoing (IBVS) method was proposed. Firstly, an eigenspace based image compression method is explored, which is chosen as the global feature transformation method. After that, the inner structure, performance and training method of T-S neural network controller are discussed respectively. Besides, the whole architecture of the TS-FNNC is investigated.

Panwar and Sukavanam in (Panwar & Sukavanam 2007) have introduced Neural Network Based Controller for Visual Servoing of Robotic Hand Eye System. For Panwar and Sukavanam, in their paper a neural network based controller for robot positioning and tracking using direct monocular visual feedback is proposed. Visual information is provided using a camera mounted on the end-effector of a robotics manipulator. A PI kinematics controller is proposed to achieve motion control objective in an image plane. A Feed forward Neural Network (FFNN) is used to compensate for the robot dynamics. The FFNN computes the required torques to drive the robot manipulator to achieve desired tracking. The stability of combined PI kinematics and FFNN computed torque is proved by Lyapunov theory. Gracia and Perez-Vidal in (Gracia & Perez-Vidal 2009), have presented a research framework through which a new control scheme for visual servoing that takes into account the delay introduced by image acquisition and image processing. The capabilities (steady-state errors, stability margins, step time response, etc.) of the proposed control scheme and of previous ones are analytically analyzed and compared. Several simulations and experimental results were provided to validate the analytical results and to illustrate the benefits of the proposed control scheme. In particular, it was shown that this novel control scheme clearly improves the performance of the previous ones.

Alessandro and researchers, as in (Alessandro et. al. 2007), in their research framework, they proposed an image-based visual servoing framework. Error signals are directly computed from image feature parameters, thus obtaining control schemes which do not need neither a 3-D model of the scene, nor a perfect knowledge of the camera calibration matrix. Value of the depth "Z" for each considered feature must be known. Thus they proposed a method to estimate on-line the value of Z for point features while the camera is moving through the scene, by using tools from nonlinear observer theory. By interpreting "Z" as a continuous unknown state with known dynamics, they build an estimator which asymptotically recovers the actual depth value for the selected feature.

In (Chen et. al. 2008), an adaptive visual servo regulation control for camera-in-hand configuration with a fixed camera extension was presented by Chen. An image-based regulation control of a robot manipulator with an uncalibrated vision system is discussed. To compensate for the unknown camera calibration parameters, a novel prediction error formulation is presented. To achieve the control objectives, a Lyapunov-based adaptive control strategy is employed. The control development for the camera-in-hand problem is presented in detail and a fixed-camera problem is included as an extension. Epipolar Geometry Toolbox as in (Gian et. al. 2004), was also created to grant MATLAB users with a broaden outline for a creation and visualization of multi-camera scenarios. In addition, to be used for the manipulation of visual information, and the visual geometry. Functions provided, for both class of vision sensing, the pinhole and panoramic, include camera assignment and visualization, computation, and estimation of epipolar geometry entities. Visual servoing has been classified as using visual data within a control loop, enabling visual-motor (hand-eye) coordination.

Image Based Visual Servoing Using Takagi-Sugeno Fuzzy Neural Network Controller has been proposed by (Miao et. al. 2007). In their study, a T-S fuzzy neural controller based IBVS method was proposed. Eigenspace based image compression method is firstly explored which is chosen as the global feature transformation method. Inner structure, performance and training method of T-S neural network controller are discussed respectively. Besides that, the whole architecture of TS-FNNC is investigated. For robotics arm visual servo, this issue has been formulated as a function of object feature Jacobian. Feature Jacobian is a complicated matrix to compute for real-time applications. For more feature points in space, the issue of computing inverse of such matrix is even more hard to achieve.

### 1.3 Chapter contribution

For robotics arm visual servo, this issue has been formulated as a function of object feature Jacobian. Feature Jacobian Matrix entries are complicated differential relations to be computed for real-time applications. For more feature points in space, the issue of computing inverse of such matrix is even more hard to achieve. In this respect, this chapter concentrates on approximating differential visual information relations relating an object movement in space to the object motion in camera space (which usually complicated relation), hence to joint space. This is known as the (object feature points). The proposed methodology will also discuss how a trained learning system can be used to achieve the needed approximation. The proposed methodology is entirely based on utilizing and merging of three MatLab Tool Boxes. Robotics Toolbox developed by Peter Corke (Corke, 2002), secondly is the Epipolar Geometry Toolbox (EGT) developed by Eleonora Alunno (Eleonora et. al. 2004), whereas the third is the ANN MatLab Tool Box.

This chapter is presenting a research framework which was oriented to develop a robotics visual servo system that relies on approximating complicated nonlinear visual servo kinematics. It concentrates on approximating differential visual changes (object features) relations relating objects movement in a world space to object motion in a camera space (usually time-consuming relations as expressed by time-varying Jacobian matrix), hence to a robotics arm joint space.

The research is also presenting how a trained Neural Network can be utilized to learn the needed approximation and inter-related mappings. The research whole concept is based on utilizing and mergence three fundamentals. The first is a robotics arm-object visual kinematics, the second is the Epipolar Geometry relating different object scenes during motion, and a learning artificial neural system. To validate the concept, the visual control loop algorithm developed by RIVES has been thus adopted to include a learning neural system. Results have indicated that, the proposed visual servoing methodology was able to produce a considerable accurate results.

## 1.4 Chapter organization

The chapter has been sub-divided into six main sections. In this respect, in section (1) we introduce the concept of robotics visual servo and related background, as related to visual servo. In section (2), we present a background and some related literatures for single scene via signal camera system. Double scene, as known as Epipolar Geometry is also presented in depth in section (3). Artificial Neural Net based **IBVS** is also presented in Section (4), whereas simulated of learning and training of an Artificial Neural Net is presented in section (5). Section (5) presents a case study and a simulation result of the proposed method, as compared to RIVES algorithm. Finally, Section (6) presents the chapter conclusions.

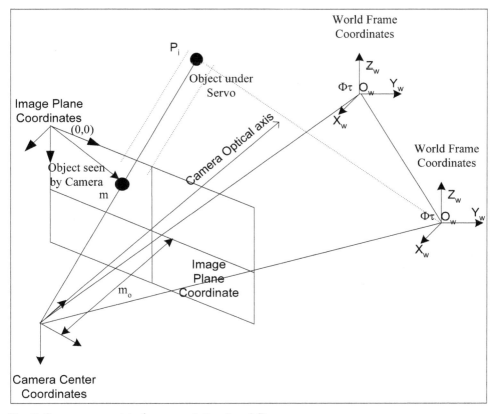

Fig. 2. Camera geometrical representation in a 3-D space

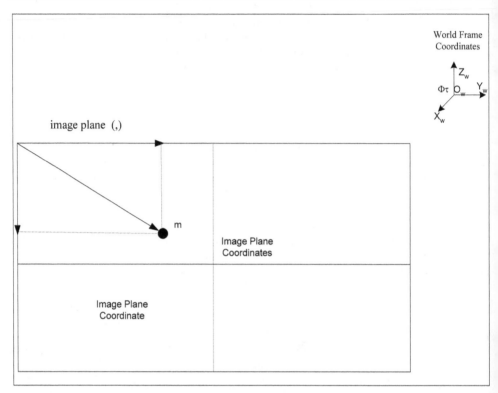

Fig. 3. Camera image plane and geometry

## 2. Single camera model: {object to camera plane kinmeatics}

Fig. (2) shows a typical camera geometrical representation in a space. Hence, to assemble a closed loop visual servo system, a loop is to be closed around the robotics arm system. In this study, this is a PUMA-560 robotics arm, with a Pinhole camera system. The camera image plane and associated geometry is shown in Fig. (3). For analyzing closed loop visual kinematics, we shall employ a Pinhole Camera Model for capturing object features. For whole representation, details of a Pinhole camera model in terms of image plane ($\xi^a$, $\psi^a$) location are expressed in terms ($\xi$, $\psi$, and $\zeta$), as in Equ. (1). In reference to Fig. (2), we can express $\left(\xi^a, \psi^a\right)$ locations as expressed in terms $(\xi, \Psi, \xi)$:

$$
\begin{cases}
\xi^a = \phi^a\left(\dfrac{\xi}{\zeta}\right) \\[3mm]
\psi^a = \phi^a\left(\dfrac{\psi}{\zeta}\right)
\end{cases}
\tag{1}
$$

Both $\left(\xi^a, \psi^a\right)$ location over an image plane is thus calculated by Equ. (1). For thin lenses (as the Pinhole camera model), camera geometry are thus represented by, (Gian et. al. 2004) :

$$\left(\frac{1}{\phi}\right)=\left(\frac{1}{\zeta^a}-\frac{1}{\zeta}\right) \tag{2}$$

In reference to (Gian et. al. 2004), using Craig Notation, denotes coordinate of point P in frame B. For translation case :

$$^B P = {}^A P + {}^B O_A \tag{3}$$

$^B O_A$ is a coordinate of the origin $O_A$ of frame "A" in a new coordinate system "B". Rotations are thus expressed :

$$^B_A R = \begin{pmatrix} {}^B i_A & {}^B j_A & {}^B k_A \end{pmatrix} = \begin{pmatrix} {}^A i_B^T \\ {}^A j_B^T \\ {}^A k_B^T \end{pmatrix} \tag{4}$$

In Equ (4), $^B i_A$ is a frame axis coordinate of "A" in another coordinate "B". In this respect, for rigid transformation we have:

$$^B P = {}^B_A R \, {}^A P$$

$$^B P = {}^B_A R \, {}^A P + {}^B O_A \tag{5}$$

For more than single consecutive rigid transformations, (for example to frame "C"), i.e. form frames $A \to B \to C$, coordinate of point P in frame "C" can hence be expressed by:

$$^B P = {}^B_A R \left( {}^B_A R \, {}^A P + {}^B O_A \right) + {}^C O_B$$

$$^B P = \left( {}^C_B R \, {}^B_A R \, {}^A P \right) + \left( {}^C_B R \, {}^A O_A + {}^C O_A \right) \tag{6}$$

For homogeneous coordinates, it looks very concise to express $^B$P as :

$$\begin{pmatrix} {}^B P \\ 1 \end{pmatrix} = \begin{pmatrix} {}^B_A R & {}^B O_A \\ O^T & 1 \end{pmatrix} \begin{pmatrix} {}^A P \\ 1 \end{pmatrix} \tag{7}$$

$$\begin{pmatrix} {}^C P \\ 1 \end{pmatrix} = \begin{pmatrix} {}^C_B R & {}^C O_B \\ O^T & 1 \end{pmatrix} \begin{pmatrix} {}^B P \\ 1 \end{pmatrix} \tag{8}$$

$$\begin{pmatrix} {}^C P \\ 1 \end{pmatrix} = \begin{pmatrix} {}^C_B R & {}^C O_B \\ O^T & 1 \end{pmatrix} \begin{pmatrix} {}^B_A R & {}^B O_A \\ O^T & 1 \end{pmatrix} \begin{pmatrix} {}^A P \\ 1 \end{pmatrix} \tag{9}$$

We could express elements of a transformation ( $\Gamma$ ) by writing :

$$\Gamma = \begin{pmatrix} \sigma_{11} & \sigma_{12} & \sigma_{13} & \sigma_{14} \\ \sigma_{21} & \sigma_{22} & \sigma_{23} & \sigma_{24} \\ \sigma_{31} & \sigma_{32} & \sigma_{33} & \sigma_{34} \\ \sigma_{41} & \sigma_{42} & \sigma_{43} & \sigma_{44} \end{pmatrix} \qquad \Gamma = \begin{pmatrix} A & \Omega \\ O^T & 1 \end{pmatrix} \tag{10}$$

as becoming offline transformation. If $(A = R)$, i.e., a rotation matrix $(\Gamma)$, once $R^T R = I$, then :

$$\Gamma = \begin{pmatrix} R & \Omega \\ O^T & 1 \end{pmatrix} \tag{11}$$

Euclidean transformation preserves parallel lines and angles, on the contrary, affine preserves parallel lines but not angles, Introducing a normalized image plane located at focal length $\phi = 1$ . For this normalized image plane, pinhole (C) is mapped into the origin of an image plane using:

$$\hat{P} = (\hat{u} \quad \hat{v})^T \tag{12}$$

$\hat{C}$ and P are mapped to :
$$\hat{P} = \begin{pmatrix} \hat{u} \\ \hat{u} \\ 1 \end{pmatrix} = \frac{1}{\zeta^C}(I \quad 0)\begin{pmatrix} P^C \\ 1 \end{pmatrix} \tag{13}$$

$$\hat{P} = \frac{1}{\zeta^C}(I \quad 0)\begin{pmatrix} \xi^c \\ \psi^c \\ \zeta^c \\ 1 \end{pmatrix} \tag{14}$$

we also have :
$$\begin{pmatrix} u \\ v \\ 1 \end{pmatrix} = \frac{1}{\zeta^C}\begin{pmatrix} k\phi & 0 & a_0 \\ 0 & k\phi & v_0 \\ 0 & 0 & 1 \end{pmatrix}\begin{pmatrix} \xi^c \\ \psi^c \\ \zeta^c \end{pmatrix}$$

$$\begin{pmatrix} u \\ v \\ 1 \end{pmatrix} = \left(\frac{1}{\zeta^C}\right)\begin{pmatrix} k\phi & 0 & u_0 \\ 0 & k\phi & v_0 \\ 0 & 0 & 1 \end{pmatrix}(I \quad 0)\begin{pmatrix} \xi^c \\ \psi^c \\ \zeta^c \\ 1 \end{pmatrix} \tag{15}$$

Now once $\kappa = k\phi$ and $\beta = L\phi$ , then we identify these parameters $\kappa, \beta, u_0, v_0$ as intrinsic camera parameters. In fact, they do present an inner camera imaging parameters. In matrix notation, this can be expressed as :

$$\begin{pmatrix} u \\ v \\ 1 \end{pmatrix} = \left(\frac{1}{\varsigma^c}\right) \begin{pmatrix} k\phi & 0 & u_0 \\ 0 & k\phi & v_0 \\ 0 & 0 & 1 \end{pmatrix} \begin{pmatrix} \xi^c \\ \psi^c \\ \varsigma^c \\ 1 \end{pmatrix}$$

$$\begin{pmatrix} u \\ v \\ 1 \end{pmatrix} = \left(\frac{1}{\varsigma^c}\right) \begin{pmatrix} \kappa & 0 & u_0 & 0 \\ 0 & \beta & v_0 & 0 \\ 0 & 0 & 1 & 0 \end{pmatrix} \begin{pmatrix} R & \Omega \\ O^T & 0 \end{pmatrix} \begin{pmatrix} \xi^c \\ \psi^c \\ \varsigma^c \\ 1 \end{pmatrix} \tag{16}$$

Both $(R)$ and $(\Omega)$ extrinsic camera parameters, do represent coordinate transformation between camera coordinate system and world coordinate system. Hence, any $(u,v)$ point in camera image plan is evaluated via the following relation:

$$\begin{pmatrix} u \\ v \\ 1 \end{pmatrix} = \left(\frac{1}{\varsigma^c}\right) M_1 M_2 p^w \qquad \begin{pmatrix} u \\ v \\ 1 \end{pmatrix} = \frac{1}{\varsigma^c} M p^w \tag{17}$$

Here $(M)$ in Equ (17) is referred to as a Camera Projection Matrix. We are given (1) a calibration rig, i.e., a reference object, to provide the world coordinate system, and (2) an image of the reference object. The problem is to solve (a) the projection matrix, and (b) the intrinsic and extrinsic parameters.

## 2.1 Computing a projection matrix
In a mathematical sense, we are given $\left(\xi_i^w \quad \psi_i^w \quad \varsigma_i^w\right)$ and $\left(u_i \quad v_i\right)^T$ for $i = (1 \ldots\ldots n)$, we want to solve for $M_1$ and $M_2$:

$$\begin{pmatrix} u_i \\ v_i \\ 1 \end{pmatrix} = \frac{1}{\varsigma^c} M_1 M_2 \begin{pmatrix} \xi_i^w \\ \psi_i^w \\ \varsigma_i^w \\ 1 \end{pmatrix} \tag{18}$$

$$\begin{pmatrix} u_i \\ v_i \\ 1 \end{pmatrix} = \frac{1}{\varsigma^c} M \begin{pmatrix} \xi_i^w \\ \psi_i^w \\ \varsigma_i^w \\ 1 \end{pmatrix}$$

$$\zeta_i^c = \begin{bmatrix} u_i \\ v_i \\ 1 \end{bmatrix} = \begin{bmatrix} m_{11} & m_{12} & m_{13} & m_{14} \\ m_{21} & m_{22} & m_{23} & m_{24} \\ m_{31} & m_{32} & m_{33} & m_{34} \end{bmatrix} \begin{bmatrix} \xi_i^w \\ \psi_i^w \\ \zeta_i^w \\ 1 \end{bmatrix} \tag{19}$$

$$\zeta_i^c u_i = m_{11}\xi_i^w + m_{12}\psi_i^w + m_{13}\zeta_i^w + m_{14}$$

$$\zeta_i^c u_i = m_{21}\xi_i^w + m_{22}\psi_i^w + m_{23}\zeta_i^w + m_{24} \tag{20}$$

$$\zeta_i^c u_i = m_{31}\xi_i^w + m_{32}\psi_i^w + m_{33}\zeta_i^w + m_{34}$$

$$\zeta_i^c = \begin{bmatrix} u_i \\ v_i \\ 1 \end{bmatrix} = \begin{bmatrix} \sigma_{11} & \sigma_{12} & \sigma_{13} & \sigma_{14} \\ \sigma_{21} & \sigma_{22} & \sigma_{23} & \sigma_{24} \\ \sigma_{31} & \sigma_{32} & \sigma_{33} & \sigma_{34} \end{bmatrix} \begin{bmatrix} \xi_i^w \\ \psi_i^w \\ \zeta_i^w \\ 1 \end{bmatrix} \tag{21}$$

$$\zeta_i^c u_i = \sigma_{21}\xi_i^w + \sigma_{22}\psi_i^w + \sigma_{23}\zeta_i^w + \sigma_{24} \tag{22}$$

$$\zeta_i^c u_i = \sigma_{31}\xi_i^w + \sigma_{32}\psi_i^w + \sigma_{33}\zeta_i^w + \sigma_{34}$$

Obviously, we can let $\sigma_{34} = 1$. This will result in the projection matrix is scaled by $\sigma_{34}$. Once $KM = U$, $K \in \Re^{2n \times 11}$ is a matrix, a 11-D vector, and $U \in \Re^{2n-D}$ vector. A least square solution of equation $KM = U$ is thus expressed by:

$$M = K^+ U$$

$$M = K^T K^{-1} K^T U \tag{23}$$

$K^+$ is the pseudo inverse of matrix $K$, and $m$ and $m_{34}$ consist of the projection matrix $M$. We now turn to double scene analysis.

## 3. Double camera scene {epipolar geometry analysis}

In this section, we shall consider an image resulting from two camera views. For two perspective views of the same scene taken from two separate viewpoints $O_1$ and $O_2$, this is illustrated in Fig. (3). Also we shall assume that ($m_1$ and $m_2$) are representing two separate points in two views. In other words, perspective projection through $O_1$ and $O_2$, of the same point $X_w$, in both image planes $\Lambda_1$ and $\Lambda_2$. In addition, by letting ($c_1$) and ($c_2$) be the optical centers of two scene, the projection $E_1$ ($E_2$) of one camera center $O_1$ ($O_2$) onto the image plane of the other camera frame $\Lambda_2$ ($\Lambda_1$) is the epipole geometry.

It is also possible to express such an epipole geometry in homogeneous coordinates in terms $\tilde{E}_1$ and $\tilde{E}_2$ :

$$\tilde{E}_1 = \begin{pmatrix} E_{1x} & E_{1y} & 1 \end{pmatrix}^T \text{ and } \tilde{E}_2 = \begin{pmatrix} E_{2x} & E_{2y} & 1 \end{pmatrix}^T \tag{24}$$

One of the main parameters of an epipolar geometry is the fundamental Matrix H (which is $\Re \in 3\times 3$). In reference to the H matrix, it conveys most of the information about the relative position and orientation ( t,R ) between the two different views. Moreover, the fundamental matrix H algebraically relates corresponding points in the two images through the Epipolar Constraint. For instant, let the case of two views of the same 3-D point $X_w$, both characterized by their relative position and orientation ( t,R ) and the internal, hence H is evaluated in terms of $K_1$ and $K_2$, representing extrinsic camera parameters, (Gian et al., 2004) :

$$H = K_2^{-T} (t)_x R K_1^{-1} \tag{25}$$

In such a similar case, a 3-D point ( $X_w$ ) is projected onto two image planes, to points ( $m_2$ ) and ( $m_1$ ), as to constitute a conjugate pair. Given a point ( $m_1$ ) in left image plane, its conjugate point in the right image is constrained to lie on the epipolar line of ( $m_1$ ). The line is considered as the projection through $C_2$ of optical ray of $m_1$. All epipolar lines in one image plane pass through an epipole point.

This is a projection of conjugate optical centre: $\tilde{E}_1 = \tilde{P}_2 \begin{pmatrix} c_1 \\ 1 \end{pmatrix}$ . Parametric equation of epipolar

line of $\tilde{m}_1$ gives $\tilde{m}_2^T = \tilde{E}_2 + \lambda P_2 P_1^{-1} \tilde{m}_1$ . In image coordinates this can be expressed as:

$$(u) = (m_2)_1 = \left( \frac{(\tilde{e}_2)_1 + \lambda(\tilde{v})_1}{(\tilde{e}_2)_3 + \lambda(\tilde{v})_3} \right) \tag{26}$$

$$(v) = (m_2)_2 = \left( \frac{(\tilde{e}_2)_2 + \lambda(\tilde{v})_2}{(\tilde{e}_2)_3 + \lambda(\tilde{v})_3} \right) \tag{27}$$

here $\tilde{v} = P_2 P_2^{-1} \tilde{m}_1$ is a projection operator extracting the ($i$th ) component from a vector.
When ( $C_1$ ) is in the focal plane of right camera, the right epipole is an infinity, and the epipolar lines form a bundle of parallel lines in the right image. Direction of each epipolar line is evaluated by derivative of parametric equations listed above with respect to ( $\lambda$ ) :

$$\left( \frac{du}{d\lambda} \right) = \left( \frac{[\tilde{v}]_1 [\tilde{e}_2]_3 - [\tilde{v}]_3 [\tilde{e}_2]_1}{([\tilde{e}_2]_3 + \lambda[\tilde{v}]_3)^2} \right) \tag{28}$$

$$\left( \frac{dv}{d\lambda} \right) = \left( \frac{[\tilde{v}]_2 [\tilde{e}_2]_3 - [\tilde{v}]_3 [\tilde{e}_2]_2}{([\tilde{e}_2]_3 + \lambda[\tilde{v}]_3)^2} \right) \tag{29}$$

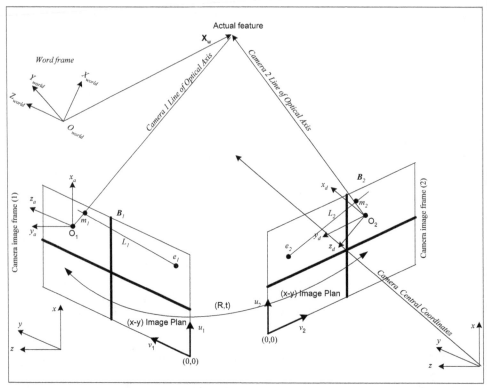

Fig 4. Camera image frame (Epipolar Geometry)

The epipole is projected to infinity once $\left(\tilde{E}_2\right)_3 = 0$. In such a case, direction of the epipolar lines in right image doesn't depend on any more. All epipolar lines becomes parallel to vector $\left(\left[\tilde{E}_2\right]_1 \; \left[\tilde{E}_2\right]_2\right)^T$. A very special occurrence is once both epipoles are at infinity. This happens once a line containing $(C_1)$ and $(C_2)$, the baseline, is contained in both focal planes, or the retinal planes are parallel and horizontal in each image as in Fig. (4). The right pictures plot the epipolar lines corresponding to the point marked in the left pictures. This procedure is called rectification. If cameras share the same focal plane the common retinal plane is constrained to be parallel to the baseline and epipolar lines are parallel.

## 4. Neural net based Image - Based Visual Servo control (ANN-IBVS)

Over the last section we have focused more in single and double camera scenes, i.e. representing the robot arm visual sensory input. In this section, we shall focus on "Image-Based Visual Servo" (IBVS) which uses locations of object features on image planes (epipolar) for direct visual feedback. For instant, while reconsidering Fig. (1), it is desired to move a robotics arm in such away that camera's view changes from ( an initial) to (final) view, and feature vector from ($\phi_0$) to ($\phi_d$). Here ($\phi_0$) may comprise coordinates of vertices, or areas of the object to be tracked. Implicit in ($\phi_d$) is the robot is normal to, and centered over features of an object, at a desired distance. Elements of the task are thus specified in image space. For a robotics system with an end-effector mounted camera, viewpoint and features are functions of relative pose of the camera to the target, ($^{c}\xi_{t}$). Such function is usually nonlinear and cross-coupled. A motion of end-effectors DOF results in complex motion of many features. For instant, a camera rotation can cause features to translate horizontally and vertically on the same image plane, as related via the following relationship :

$$\phi = f\left(^{c}\xi_{t}\right) \tag{30}$$

Equ (30) is to be linearized. This is to be achieved around an operating point:

$$\delta\phi = {}^{f}J_{c}\left(^{c}x_{t}\right)\delta^{c}x_{t} \tag{31}$$

$${}^{f}J_{c}\left(^{c}x_{t}\right) = \left(\frac{\partial\phi}{\partial^{c}x_{t}}\right) \tag{32}$$

In Equ (32), ${}^{f}J_{c}\left(^{c}x_{t}\right)$ is the Jacobian matrix, relating rate of change in robot arm pose to rate of change in feature space. Variously, this Jacobian is referred to as the feature Jacobian, image Jacobian, feature sensitivity matrix, or interaction matrix. Assume that the Jacobian is square and non-singular, then:

$${}^{c}\dot{x}_{t} = {}^{f}J_{c}\left(^{c}x_{t}\right)^{-1}\dot{f} \tag{33}$$

from which a control law can be expressed by :

$${}^{c}\dot{x}_{t} = (K)\,{}^{f}J_{c}\left(^{c}x_{t}\right)^{-1}\left(f_{d} - f(t)\right) \tag{34}$$

will tend to move the robotics arm towards desired feature vector. In Equ (34), $K^{f}$ is a diagonal gain matrix, and (t) indicates a time varying quantity. Object posture rates $^{c}\dot{x}_{t}$ is converted to robot end-effector rates. A Jacobian , ${}^{f}J_{c}(^{c}x_{t})$ as derived from relative pose between the end-effecter and camera, $(^{c}x_{t})$ is used for that purpose. In this respect, a technique to determine a transformation between a robot's end-effector and the camera frame is given by Lenz and Tsai, as in (Lenz & Tsai. 1988). In a similar approach, an end-effector rates may be converted to manipulator joint rates using the manipulator's Jacobian (Croke, 1994), as follows:

$$\dot{\theta}_{t} = {}^{t6}J_{\theta}^{-1}(\theta)\,{}^{t6}\dot{x}_{c} \tag{35}$$

$\dot{\theta}_{t}$ represents the robot joint space rate. A complete closed loop equation can then be given by:

$$\dot{\theta}_{t} = K\,{}^{t6}J_{\theta}^{-1}(\theta)\,{}^{t6}J_{\theta}^{f}J_{c}^{-1}\,{}^{c}x_{t}\left(f_{d} - f(t)\right) \tag{36}$$

For achieving this task, an analytical expression of the error function is given by :

$$\phi = Z^+ \phi_1 + \gamma \left( I_6 - Z^+ Z \right) \frac{\partial \phi_2}{\partial X} \tag{37}$$

Here, $\gamma \in \Re^+$ and $Z^+$ is pseudo inverse of the matrix $Z$, $Z \in \Re^{m \times n} = \Re(Z^T) = \Re(J_1^T)$ and $J$ is the Jacobian matrix of task function as $J = \left( \frac{\partial \phi}{\partial X} \right)$. Due to modeling errors, such a closed-loop system is relatively robust in a possible presence of image distortions and kinematics parameter variations of the Puma 560 kinematics. A number of researchers also have demonstrated good results in using this image-based approach for visual servoing. It is always reported that, the significant problem is computing or estimating the feature Jacobian, where a variety of approaches have been used (Croke, 1994). The proposed IBVS structure of Weiss (Weiss et. al., 1987 and Craig, 2004), controls robot joint angles directly using measured image features. Non-linearities include manipulator kinematics and dynamics as well as the perspective imaging model. Adaptive control was also proposed, since ${}^f J_\theta^{-1}({}^c \theta)$, is pose dependent, (Craig, 2004). In this study, changing relationship between robot posture and image feature change is learned during a motion via a learning neural system. The learning neural system accepts a weighted set of inputs (stimulus) and responds.

## 4.1 Visual mapping: Nonlinear function approximation ANN mapping

A layered feed-forward network consists of a certain number of layers, and each layer contains a certain number of units. There is an input layer, an output layer, and one or more hidden layers between the input and the output layer. Each unit receives its inputs directly from the previous layer (except for input units) and sends its output directly to units in the next layer. Unlike the Recurrent network, which contains feedback information, there are no connections from any of the units to the inputs of the previous layers nor to other units in the same layer, nor to units more than one layer ahead. Every unit only acts as an input to the immediate next layer. Obviously, this class of networks is easier to analyze theoretically than other general topologies because their outputs can be represented with explicit functions of the inputs and the weights.

In this research we focused on the use of Back-Propagation Algorithm as a learning method, where all associated mathematical used formulae are in reference to Fig. (5). The figure depicts a multi-layer artificial neural net (a four layer) being connected to form the entire network which learns using the Back-propagation learning algorithm. To train the network and measure how well it performs, an objective function must be defined to provide an unambiguous numerical rating of system performance. Selection of the objective function is very important because the function represents the design goals and decides what training algorithm can be taken. For this research frame work, a few basic cost functions have been investigated, where the sum of squares error function was used as defined by Equ. (38):

$$E = \frac{1}{NP} \sum_{p=1}^{P} \sum_{i=1}^{N} \left( t_{pi} - y_{pi} \right)^2 \tag{38}$$

where p indexes the patterns in the training set, $i$ indexes the output nodes, and $t_{pi}$ and $y_{pi}$ are, respectively, the target hand joint space position and actual network output for the $i^{th}$ output unit on the $p^{th}$ pattern. An illustration of the layered network with an input layer, two hidden layers, and an output layer is shown in Fig. (5). In this network there are $i$

inputs, ($m$) hidden units, and ($n$) output units. The output of the $j^{th}$ hidden unit is obtained by first forming a weighted linear combination of the ($i$) input values, then adding a bias:

$$a_j = \left( \sum_{i=1}^{I} w_{ji}^1 x_i + w_{j0}^1 \right) \tag{39}$$

where $w_{ji}^{(1)}$ is a weight from input ($i$) to hidden unit ($j$) in the first layer. $w_{j0}^{(1)}$ is a bias for hidden unit $j$. If we are considering a bias term as being weights from an extra input $x_0 = 1$, Equ. (39) can be rewritten to the form of:

$$a_j = \left( \sum_{i=0}^{I} w_{ji}^1 x_i \right) \tag{40}$$

The activation of hidden unit $j$ then can be obtained by transforming the linear sum using a *nonlinear activation function* $g(x)$:

$$h_j = g\left( a_j \right) \tag{41}$$

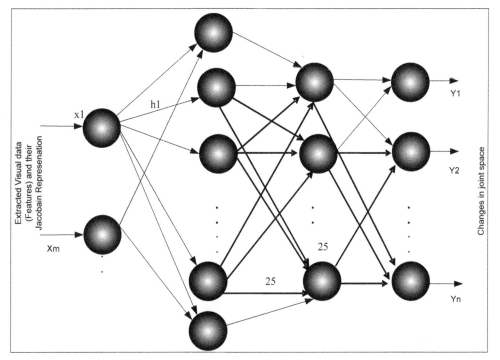

Fig. 5. Employed four layers artificial neural system

Outputs of the neural net is obtained by transforming the activation of the hidden units using a second layer of processing units. For each output unit k, first we get the linear combination of the output of the hidden units, as in Equ. (42):

$$a_k = \left( \sum_{j=1}^{m} w^2_{kj} h_j + w^2_{k0} \right) \qquad (42)$$

absorbing the bias and expressing the above equation to:

$$a_k = \left( \sum_{j=0}^{m} w^2_{kj} h_j \right) \qquad (43)$$

Applying the activation function $g_2(x)$ to Equ. (43), we can therefore get the $k^{th}$ output :

$$y_k = g_2(a_k) \qquad (44)$$

Combining Equ. (40), Equ. (41), Equ. (43) and Equ. (44), we get a complete representation of the network as:

$$y_k = g_2 \left( \sum_{j=0}^{m} w^2_{kj} g \left( \sum_{i=0}^{l} w^2_{ji} x_i \right) \right) \qquad (45)$$

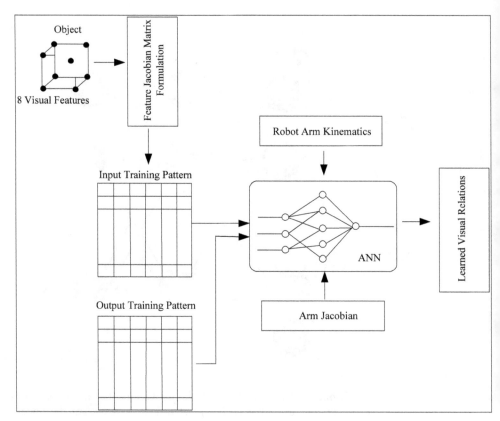

Fig. 6. Features based data gathering: Training patterns generations

The network of Fig. (5) is a synthesized ANN network with two hidden layers, which can be extended to have extra hidden layers easily, as long as we make the above transformation further. Input units do transform neural network signals to the next processing nodes. They are hypothetical units that produce outputs equal to their supposed inputs, hence no processing is done by these input units. Through this approach, the error of the network is propagated backward recursively through the entire network and all of the weights are adjusted so as to minimize the overall network error. The block diagram of the used learning neural network is illustrated in Fig. (6). The network learns the relationship between the previous changes in the joint angles $\Delta\Theta_{k-1}$, changes in the object posture $\Delta u_a^c$, and changes joint angles $\Delta\Theta_k$. This is done by executing some random displacements from the desired object position and orientation. The hand fingers is set up in the desired position and orientation to the object. Different Cartesian based trajectories are then defined and the inverse Jacobian were used to compute the associated joints displacement $\Theta_h(k)$. Different object postures with joint positions and differential changes in joint positions are the input-output patterns for training the employed neural network. During the learning epoch, weights of connections of neurons and biases are updated and changed, in such away that errors decrease to a value close to zero, which resulted in the learning curve that minimizes the defined objective function shown as will be further discussed later. It should be mentioned at this stage that the training process has indeed consumed nearly up to three hours, this is due to the large mount of training patterns to be presented to the neural network.

## 4.2 Artificial neural networks mapping: A biological inspiration

Animals are able to respond adaptively to changes in their external and internal environment and surroundings, and they use their nervous system to perform these behaviours. An appropriate model/simulation of a nervous system should be able to produce similar responses and behaviours in artificial systems. A nervous system is built by relatively simple; units, the neurons, so copying their behaviour and functionality should be the solution, (Pellionisz, 1989). In reality, human brain is a part of the central nervous system, it contains of the order of $(10^{+10})$ neurons. Each can activate in approximately 5ms and connects to the order of $(10^{+4})$ other neurons giving $(10^{+14})$ connections, (Shields & Casey, 2008). In reality, a typical neural net (with neurons) is shown in Fig. (5), it does resemble actual biological neuron, as they are made of:

- *Synapses*: Gap between adjacent neurons across which chemical signals are transmitted: (known as the input)
- *Dendrites*: Receive synaptic contacts from other neurons
- *Cell body /soma*: Metabolic centre of the neuron: processing
- *Axon*: Long narrow process that extends from body: (known as the output)

By emulation, ANN information transmission happens at the synapses, as shown in Fig. (5). Spikes travelling along the axon of the pre-synaptic neuron trigger the release of neurotransmitter substances at the synapse. The neurotransmitters cause excitation or inhibition in the dendrite of the post-synaptic neuron. The integration of the excitatory and inhibitory signals may produce spikes in the post-synaptic neuron. The contribution of the signals depends on the strength of the synaptic connection (Pellionisz, 1989). An Artificial Neural Network (ANN) is an information processing paradigm that is inspired by the way biological nervous systems, such as the brain, process information. The key element of this paradigm is the novel structure of the information processing system. It is composed of a large number of highly interconnected processing elements (neurons) working in unison to solve specific problems. ANN system, like people, learn by example.

An ANN is configured for a specific application, such as pattern recognition or data classification, through a learning process. Learning in biological systems involves adjustments to the synaptic connections that exist between the neurons. This is true of ANN system as well (Aleksander & Morton, 1995). The four-layer feed-forward neural network with ($n$) input units, ($m$) output units and N units in the hidden layer, already shown in the Fig. (5), and as will be further discussed later. In reality, Fig. (5). exposes only one possible neural network architecture that will serve the purpose. In reference to the Fig. (5), every node is designed in such away to mimic its biological counterpart, the neuron. Interconnection of different neurons forms an entire grid of the used ANN that have the ability to learn and approximate the nonlinear visual kinematics relations. The used learning neural system composes of four layers. The {input}, {output} layers, and two {hidden layers}. If we denote ($^{w}v_c$) and ($^{w}\omega_c$) as the camera's linear and angular velocities with respect to the robot frame respectively, motion of the image feature point as a function of the camera velocity is obtained through the following matrix relation:

$$\dot{\gamma}=-\left(\frac{\alpha\lambda}{^{c}p_{c}}\right)\begin{pmatrix}0 & 0 & \dfrac{^{c}p_{x}}{^{c}p_{z}} & \dfrac{^{c}p_{x}\,^{c}p_{x}}{^{c}p_{z}} & -\dfrac{^{c}p_{x}\,^{c}p_{x}}{^{c}p_{z}} & ^{c}p_{x} \\ 1 & -1 & \dfrac{^{c}p_{y}}{^{c}p_{z}} & \dfrac{^{c}p_{x}\,^{c}p_{x}}{^{c}p_{z}} & -\dfrac{^{c}p_{x}\,^{c}p_{x}}{^{c}p_{z}} & -^{c}p_{x}\end{pmatrix}\begin{pmatrix}^{c}R_{w} & 0 \\ 0 & ^{c}R_{w}\end{pmatrix}\begin{pmatrix}^{w}v_{c} \\ ^{w}\omega_{c}\end{pmatrix} \qquad (46)$$

Instead of using coordinates ($^{x}P_c$) and ($^{y}P_c$) for the object feature described in camera coordinate frame, which are a priori unknown, it is usual to replace them by coordinates ($u$) and ($v$) of the projection of such a feature point onto the image frame, as shown in Fig. (7).

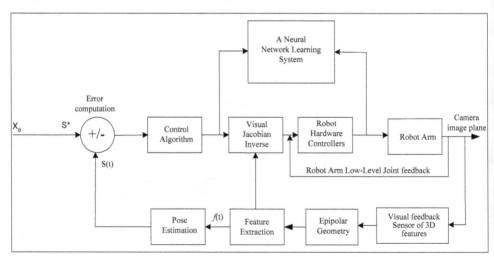

Fig. 7. Neural net based visual servo system

# 5. Simulation case study: Visual servoing with pin-hole camera for 6-DOF PUMA robotics arm

Visual servoing using a "pin-hole" camera for a 6-DOF robotics arm is simulated here. The system under study is a (PUMA) and integrated with a camera and ANN. Simulation block is shown Fig. (7). Over simulation, the task has been performed using 6-DOF-PUMA manipulator with 6 revolute joints and a camera that can provide position information of the robot gripper tip and a target (object) in robot workplace. The robot dynamics and direct kinematics are expressed by a set of equations of PUMA-560 robotics system, as documented by Craig, (Craig, 2004). Kinematics and dynamics equations are already well known in the literature, therefore. For a purpose of comparison, the used example is based on visual servoing system developed by RIVES, as in (Eleonora, 2004). The robotics arm system are has been servoing to follow an object that is moving in a 3-D working space. Object has been characterized by some like (8-features) marks, this has resulted in 24, $\Re^{8 \times 3}$ size, feature Jacobian matrix. This is visually shown in Fig. (7). An object 8-features will be mapped to the movement of the object in the camera image plane through defined geometries. Changes in features points, and the differential changes in robot arm, constitute the data that will be used for training the ANN. The employed ANN architecture has already been discussed and presented in Fig. (5).

## 5.1 Training phase: visual training patterns generation

The foremost ambition of this visual servoing is to drive a 6 DOF robot arm, as simulated with Robot Toolbox (Corke , 2002), and equipped with a pin-hole camera, as simulated with Epipolar Geometry Toolbox, EGT (Gian et al., 2004), from a starting configuration toward a desired one using only image data provided during the robot motion. For the purpose of setting up the proposed method, RIVES algorithm has been run a number of time before hand. In each case, the arm was servoing with different object posture and a desired location in the working space. The EGT function to estimate the fundamental matrix H , given $U_1$ and $U_2$, for both scenes in which $U_1$ and $U_2$ are defined in terms of eight ($\xi$ , $\psi$ , $\zeta$) feature points:

$$U_1 = \begin{pmatrix} \xi_1^1 & \xi_1^2 & \cdots & \xi_1^8 \\ \psi_1^1 & \psi_1^2 & \cdots & \xi_1^8 \\ \zeta_1^1 & \zeta_1^2 & \cdots & \zeta_1^8 \end{pmatrix}$$

and

$$U_2 = \begin{pmatrix} \xi_2^1 & \xi_2^2 & \cdots & \xi_2^8 \\ \psi_2^1 & \psi_2^2 & \cdots & \xi_2^8 \\ \zeta_2^1 & \zeta_2^2 & \cdots & \zeta_2^8 \end{pmatrix}$$

(47)

Large training patterns have been gathered and classified, therefore. Gathered patterns at various loop locations gave an inspiration to a feasible size of learning neural system. Four layers artificial neural system has been found a feasible architecture for that purpose. The net maps 24 (3×8 feature points) inputs characterizing object cartesian feature position and arm joint positions into the (six) differential changes in arm joints positions. The network is presented with some arm motion in various directions. Once the neural system has learned with presented patterns and required mapping, it is ready to be employed in the visual servo controller. Trained neural net was able to map nonlinear relations relating object movement to differentional changes in arm joint space. Object path of motion was defined and simulated via RIVES Algorithm, as given in (Gian et al., 2004), after such large number of running and patterns, it was apparent that the learning neural system was able to capture such nonlinear relations.

## 5.2 The execution phase

Execution starts primary while employing learned neural system within the robotics dynamic controller (which is mainly dependent on visual feature Jacobian). In reference to Fig. (7), visual servoing dictates the visual features extraction block. That was achieved by the use of the Epipolar Toolbox. For assessing the proposed visual servo algorithm, simulation of full arm dynamics has been achieved using kinematics and dynamic models for the Puma 560 arm. Robot Toolbox has been used for that purpose. In this respect, also Fig. (8) shows an "aerial view" of actual object "initial" posture and the "desired" posture. This is prior to visual servoing to take place. The figure also indicates some scene features. Over simulation, Fig. (9) shows an "aerial view" of the Robot arm-camera servoing, as

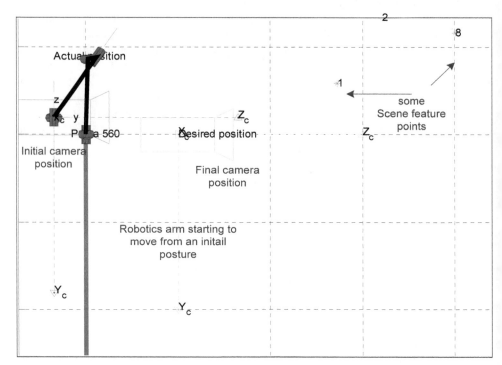

Fig. 8. Top view. Actual object position and desired position before the servoing

approaching towards a desired object posture. ANN was fed with defined patterns during arm movement. Epipolars have been used to evaluate visual features and the update during arm movement.

### 5.3 Object visual features migration

Fig. (10) shows the error between the RIVES Algorithm and the proposed ANN based visual servo for the first (60) iterations. Results suggest high accuracy of identical results, indicating that a learned neural system was able to servo the arm to desired posture. Difference in error was recorded within the range of ($4\times10^{-6}$) for specific joint angles. Fig. (11) shows migration of the eight visual features as seen over the camera image plan. Just the once the Puma robot arm was moving, concentration of features are located towards an end within camera image plane. In Fig. (12), it is shown the object six dimensional movements. They indicate that they are approaching the zero reference. As an validation of the neural network ability to servo the robotics arm toward a defined object posture, Fig. (13) show that the trained ANN visual servo controller does approach zero level of movement. This is for different training patterns and for different arm postures in the 6-dimenstional space. Finally, Fig. (14) shows the error between RIVES computed joint space values and the proposed ANN controller computed joint space values. Results indicate excellent degree of accuracy while the visual servo controller approaching the target posture with nearly zero level of erroe for different training visual servo target postures.

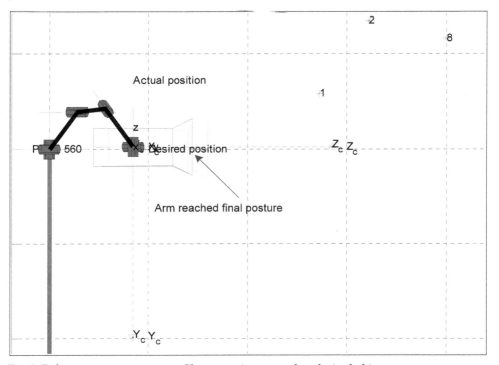

Fig. 9. Robot arm-camera system: Clear servoing towards a desired object posture

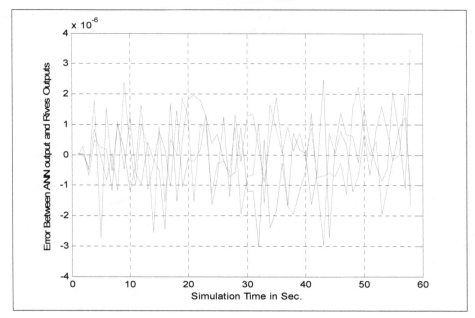

Fig. 10. Resulting errors. Use of proposed ANN based visual servo

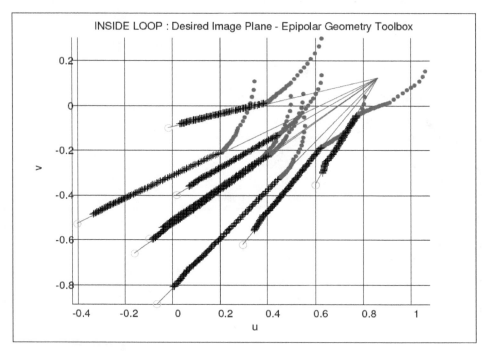

Fig. 11. Migration of eight visual features (as observed over the camera image plan)

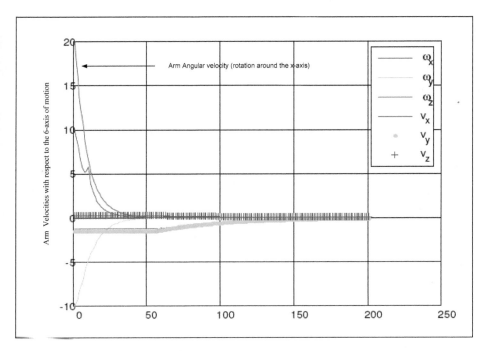

Fig. 12. Puma arm six dimensional movements

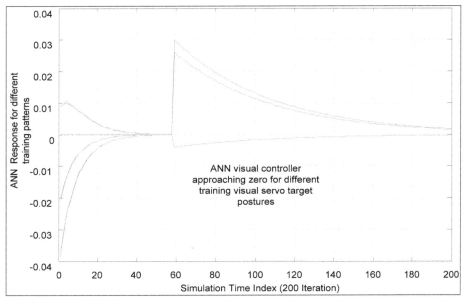

Fig. 13. ANN visual servo controller approaching zero value for different training visual servo target postures

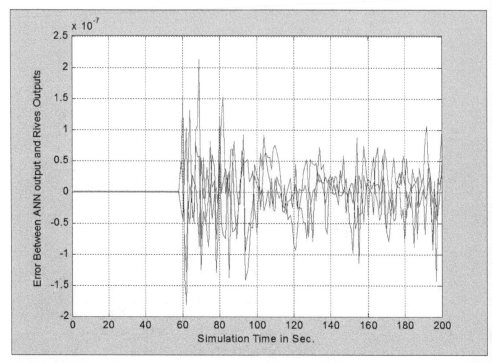

Fig. 14. Error of arm movements: ANN based controller and RIVES output difference ANN visual Servo

## 6. Conclusions

Servoing a robotics arm towards a moving object movement using visual information, is a research topic that has been presented and discussed by a number of researchers for the last twenty years. In this sense, the chapter has discussed a mechanism to learn kinematics and feature-based Jacobian relations, that are used for robotics arm visual servo system. In this respect, the concept introduced within this chapter was based on an employment and utilization of an artificial neural network system. The ANN was trained in such away to learn a mapping relating the " complicated kinematics" as relating changes in visual loop into arm joint space. Changes in a loop visual Jocobain depends heavily on a robotics arm 3-D posture, in addition, it depends on features associated with an object under visual servo (to be tracked). Results have shown that, trained neural network can be used to learn such complicated visual relations relating an object movement to an arm joint space movement. The proposed methodology has also resulted in a great deal of accuracy. The proposed methodology was applied to the well-know Image Based Visual Servoing, already discussed and presented by RIVES as documented in (Gian et al., 2004). Results have indicated a close degree of accuracy between the already published "RIVES Algorithm" results and the newly proposed "ANN Visual Servo Algorithm". This indicates ANN Visual Servo, as been based on some space learning mechanisms, can reduce the computation time.

# 7. References

Aleksander, I. & Morton, H. (1995), An Introduction to Neural Computing, *Textbook, International Edition 2nd Edition.*

Alessandro, D.; Giuseppe, L.; Paolo, O. & Giordano, R. (2007), On-Line Estimation of Feature Depth for Image-Based Visual Servoing Schemes, *IEEE International Conference on Robotics and Automation*, Italy, 1014

Chen, J.; Dawson, M.; Dixon, E. & Aman, B. (2008), Adaptive Visual Servo Regulation Control for Camera-in-Hand Configuration With a Fixed-Camera Extension, *Proceedings of the 46th IEEE Conference on Decision and Control*, CDC, 2008, pp. 2339-2344, New Orleans, LA, United States

Cisneros P. (2004), Intelligent Model Structures in Visual Servoing, *Ph.D. Thesis*, University of Manchester, Institute for Science and Technology

Corke, P. (2002), Robotics Toolbox for MatLab, *For Use with MATLAB, User Guide*, Vol. 1.

Craig, J. (2004), Introduction to Robotics: Mechanics and Control, Textbook, *International Edition, Prentice Hall*

Croke, P. (1994), High-Performance Visual Closed-Loop Robot Control, Thesis submitted in total fulfillment of the Requirements for the Degree of Doctor of Philosophy.

Eleonora, A.; Gian, M. & Domenico, P. (2004), Epipolar Geometry Toolbox, *For Use with MATLAB, User Guide*, Vol. 1.

Gian, M.; Eleonora, A. & Domenico, P. (2004), The Epipolar Geometry Toolbox (EGT) for MATLAB, *Technical Report*, 07-21-3-DII University of Siena, Siena, Italy

Gracia, L. & Perez-Vidal, C. (2009), A New Control Scheme for Visual Servoing, *International Journal of Control, Automation,and Systems*, Vol. 7, No. 5, pp. 764-776, DOI 10.1007/s12555-009-0509-9

Lenz, K. & Tsai, Y. (1988), Calibrating a Cartesian Robot with Eye-on-Hand Configuration Independent of Eye-To-Hand Relationship, *Proceedings Computer Vision and Pattern Recognition*, 1988 CVPR apos., Computer Society Conference, Vol.5, No. 9, pp. 67 – 75

Martinet, P. (2004), Applications In Visual Servoing, *IEEE-RSJ IROS'04 International Conference*, September 2004, Japan.

Miao, H.; Zengqi, S. & Fujii, M. (2007), Image Based Visual Servoing Using Takagi-Sugeno Fuzzy Neural Network Controller, *IEEE 22nd International Symposium on Intelligent Control, ISIC 2007*, Singapore, Vol. 1, No. 3, pp. 53 – 58

Miao, H.; Zengqi, S. & Masakazu, F. (2007), Image Based Visual Servoing Using Takagi-Sugeno Fuzzy Neural Network Controller, *22nd IEEE International Symposium on Intelligent Control*, Part of IEEE Multi-conference on Systems and Control Singapore, pp. 1-3

Panwar, V. & Sukavanam, N. (2007), Neural Network Based Controller for Visual Servoing of Robotic Hand Eye System, *Engineering Letters*, Vol. 14, No.1, EL_14_1_26, Advance Online Publication

Pellionisz, A. (1989), About the Geometry Intrinsic to Neural Nets, *International Joint Conference on Neural Nets*, Washington, D.C., Vol. 1, pp. 711-715

Shields, M. & Casey, C. (2008), A Theoretical Framework for Multiple Neural Network Systems, *Journal of Neurocomputing*, Vol. 71, No.7-9, pp. 1462-1476

Weiss, L.; Sanderson, A & Neuman, A. (1987), Dynamic Visual Servo Control of Robots: An adaptive Image-Based Approach , *IEEE Journal on Robotics and Automation*, Vol. 3, No.5, pp. 404–417.

# 3

# Real-Time Robotic Hand Control Using Hand Gestures

Jagdish Lal Raheja, Radhey Shyam, G. Arun Rajsekhar and P. Bhanu Prasad
*Digital Systems Group, Central Electronics Engineering Research Institute*
*(CEERI)/Council of Scientific & Industrial Research (CSIR), Pilani, Rajasthan*
*India*

## 1. Introduction

Interpretation of human gestures by a computer is used for human-machine interaction in the area of computer vision [3][28]. The main purpose of gesture recognition research is to identify a particular human gesture and convey information to the user pertaining to individual gesture. From the corpus of gestures, specific gesture of interest can be identified[30][36], and on the basis of that, specific command for execution of action can be given to robotic system[31]. Overall aim is to make the computer understand human body language [14], thereby bridging the gap between machine and human. Hand gesture recognition can be used to enhance human–computer interaction without depending on traditional input devices such as keyboard and mouse.

Hand gestures are extensively used for telerobotic control applications [20]. Robotic systems can be controlled naturally and intuitively with such telerobotic communication [38] [41]. A prominent benefit of such a system is that it presents a natural way to send geometrical information to the robot such as: left, right, etc. Robotic hand can be controlled remotely by hand gestures. Research is being carried out in this area for a long time. Several approaches have been developed for sensing hand movements and controlling robotic hand [9][13][22]. Glove based technique is a well-known means of recognizing hand gestures. It utilizes sensor attached mechanical glove devices that directly measure hand and/or arm joint angles and spatial position. Although glove-based gestural interfaces give more precision, it limits freedom as it requires users to wear cumbersome patch of devices. Jae-Ho Shin et al [14] used entropy analysis to extract hand region in complex background for hand gesture recognition system. Robot controlling is done by fusion of hand positioning and arm gestures [2][32] using data gloves. Although it gives more precision, it limits freedom due to necessity of wearing gloves. For capturing hand gestures correctly, proper light and camera angles are required. A technique to manage light source and view angles, in an efficient way for capturing hand gestures, is given in [26]. Reviews have been written on the subsets of hand postures and gesture recognition in [6]and [18]. The problem of visual hand recognition and tracking is quite challenging. Many early approaches used position markers or colored bands to make the problem of hand recognition easier, but due to their inconvenience, they cannot be considered as a natural interface for the robot control [24]. A 3D Matlab Kinematic model of a PUMA [1][4][16] robot, is used for executing actions by hand gesture[39]. It can be extended to any robotic system with a number of specific commands suitable to that system [17].

In this chapter, we have proposed a fast as well as automatic hand gesture detection and recognition system. Once a hand gesture is recognised, an appropriate command is sent to a robot. Once robot recieves a command, it does a pre-defined work and keeps doing until a new command arrives. A flowchart for overall controlling of a robot is given in figure 1.

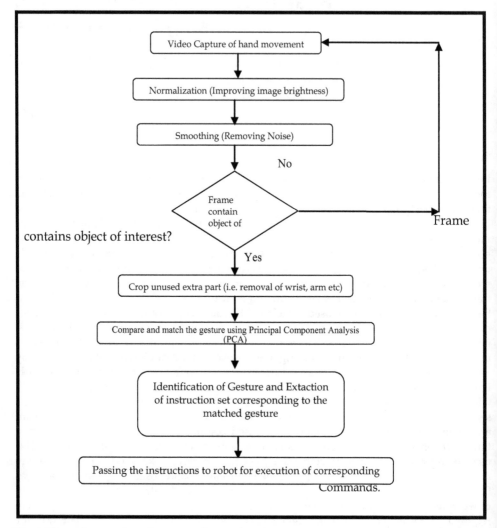

Fig. 1. Flow diagram for robot control using hand gestures

## 2. Methodology

The proposed methodology involves acquisition of live video from a camera for gesture identification. It acquire frames from live video stream in some given time interval[15]. In this case frame capture rate for gesture search is 3 frames per second. The overall proposed technique to acquire hand gestures and to control robotic system using hand gestures is divided into four subparts:

- Image acquisition to get hand gestures.
- Extracting hand gesture area from captured frame.
- Determining gesture by pattern matching using PCA algorithm.
- Generation of instructions corresponding to matched gesture, for specified robotic action.

## 2.1 Image acquisition to get hand gestures

First step is to capture the image from video stream. It is assumed that while capturing video, black background with proper light soure is used. To minimize number of light sources, arrangement of the light sources and camera is made as shown in figure 2.

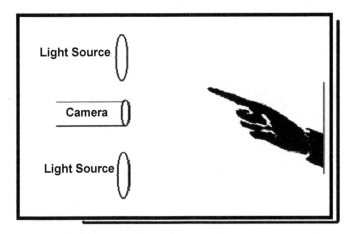

Fig. 2. Arrangement of light sources for image aquisition

From video stream one frame is captured every 1/3 of a second. Our aim is to identify the frame that contains hand gesture shown by human. For this we are searching the frame in which there is no or least movement. Required frame is identified by comparing three continuously captured frames. *Motion parameter* is determined for each frame by counting total pixels of mismatch. If *motion parameter* is less than the specified threshold value, it is considered as a frame having least movement of object i.e. the hand of the person.

Analysis of frames to find the frame of interest is carried out by converting the captured frame into a binary frame[23]. Since binary images have values of one or zero., differences of white pixels between newly captured frame and two previously captured frames are determined and they are added together to find *motion parameter*. XOR function is used to find the differences in white pixels [5][15].

If *frame1, frame2,* and, *frame3* are three matrices containing three frames captured in three continuous time slots respectively then:

$$fr1 = frame1 \ XOR \ frame3 \qquad (1)$$

$$fr2 = frame2 \ XOR \ frame \ 3 \qquad (2)$$

$$mismatch\_matrix = fr1 \ OR \ fr2 \tag{3}$$

$$\text{Sum} = \sum_{i=1}^{r} \sum_{j=1}^{c} mismatch\_matrix[i][j] \tag{4}$$

$$\text{Motion parameter} = \frac{sum}{r * c} \tag{5}$$

Here r and c are the number of rows and columns in an image frame. Threshold value is set as 0.01. i.e. if total pixels of mismatch are less than 1% of total pixels in a frame, then it is considered as frame of interest. Required frame is forwarded for further processing and this module again starts searching for next frame of interest.

## 2.2 Extracting hand gesture area from frame of interest
It may happen that the frame, as shown in fig.3, with a gesture contains extra part along with required part of hand i.e. background objects, blank spaces etc. For better results in pattern matching, unused part must be removed. Therefore hand gesture is cropped out from the obtained frame. Cropping of hand gesture from the obtained frame contains three steps:

Fig. 3. Grayscale image

First step is to convert selected frame into HSV color space. As shown in fig.4, a HSV based skin filter was applied on the RGB format image to reduce lighting effects [8][1]. The thresholds for skin filter used in the experiment were as below:

$$\left. \begin{array}{c} 0.05 < H < 0.17 \\ 0.1 < S < 0.3 \\ 0.09 < V < 0.15 \end{array} \right\} \rightarrow \tag{6}$$

The resultant image was segmented to get a binary image of hand[37]. Binary images are bi-level images where each pixel is stored as a single bit (0 or 1). Smoothening was needed, as the output image had some jagged edges as clearly visible in figure 4(c). There can be some noise in the filtered images due to false detected skin pixels or some skin color objects (like wood) in background, it can generate few unwanted spots in the output image as shown in figure 4(e).

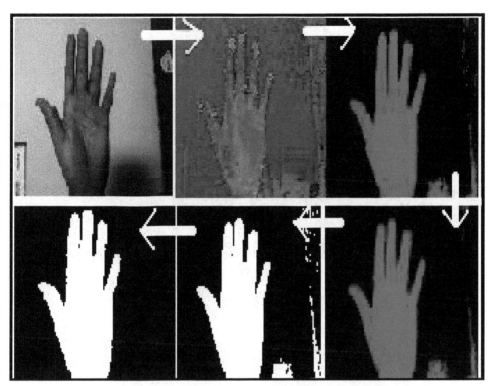

Fig. 4. Binary Image formation   (a) Target image   (b) HSV conversion   (c) Filtered image
(d) Smoothened image   (e) Binary image   (f) Biggest BLOB

To remove these errors, the biggest BLOB (Binary Linked Object) method was applied to the
noisy image [24][25]. The error free image is shown in figure 5. The only limitation of this
filter was that the BLOB for hand should be the biggest one. In this, masking background
would be illuminated, so false detection is not possible from the background [7][10].
Second step is to extract object of interest from the frame. In our case, object of interest is the
part of human hand showing gesture [29]. For this, extra part, other than the hand, is
cropped out so that pattern matching can give more accurate results [12]. For cropping extra
parts row and column numbers are determined, from where object of interest appears. This
is done by searching from each side of binary image and moving forward until white pixels
encountered are more than the offset value [25]. Experimental results show that offset value
set to 1% of total width gives better result for noise compensation. If size of selected image is
mXn then:

$$Hor\_offset = m/100 \qquad (7)$$

$$Ver\_offset = n/100 \qquad (8)$$

*Min_col* = minimum value of column number where total number of white pixels are more
than *Hor_offset*.
*Max_col* = maximum value of column number where total number of white pixels are more
than *Hor_offset*.

*Min_row*= minimum value of row number where total number of white pixels are more than *Ver_offset*.

Max_row= maximum value of row number where total number of white pixels are more than *Ver_offset*.

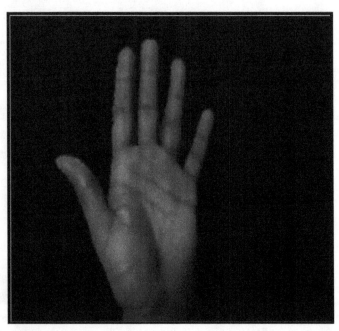

Fig. 5. Error free image

Figure 6 shows Min_row, Max_row, Min_col, and, Max_col for the selection of the boundary for cropping of hand.

Third step is to remove parts of hand not used in gesture i.e. removal of wrist, arm etc. As these extra parts are of variable length in image frame and pattern matching with gesture database gives unwanted results. Therefore, parts of hand before the wrist need to be cropped out.

Statistical analysis of hand shape shows that either we pose palm or fist, width is lowest at wrist and highest at the middle of palm. Therefore extra hand part can be cropped out from wrist by determining location where minimum width of vertical histogram is found. Figure 7.c and 7.d show global maxima and cropping points for hand gestures in figure 7.a and 7.b respectively.

Cropping point is calculated as:

*Global Maxima = column number where height of histogram is highest (i.e. column number for global maxima as shown in figure 7).*

*Cropping point = column number where height of histogram is lowest in such a way that cropping point is in between first column and column of Global Maxima*

If gesture is shown from opposite side (i.e. from other hand), then *Cropping point* is searched between column of *Global Maxima* and last column. Direction of the hand is detected using continuity analysis of object during hand gesture area determination. Continuity analysis shows that whether the object continues from the column of Global maxima to first column or to last column. i.e. whether extra hand is left side of palm or right side of palm.

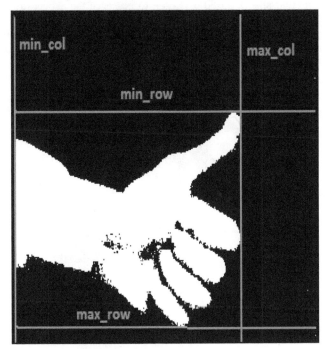

Fig. 6. Boundary selection for hand cropping

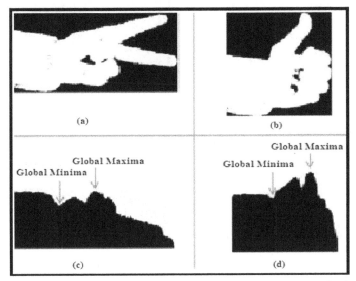

Fig. 7. Histogram showing white pixels in each column, with cropping point for hand gesture

## 2.3 Determining the gesture by resizing and pattern matching

Cropping results in a frame of 60x80 pixels, so that the gesture can be compared with the database. This particular size of 60x80 is chosen to preserve the aspect ratio of the original image, so that no distortion takes place.

Figure 8.b shows cropped hand gesture from figure 8.a. and figure 8.c. shows the result of scaling and fitting 8.b within 60X80 pixels. Figure 8.d represents final cropped and resized gesture of desired dimension.

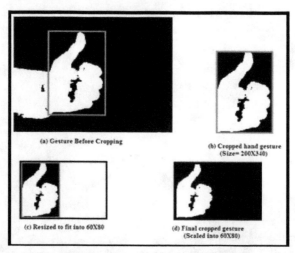

Fig. 8. Scaling cropped hand gesture to fit into desired dimension

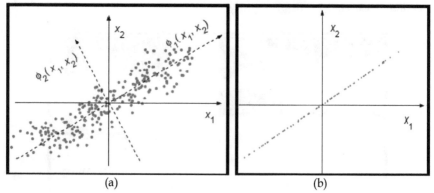

Fig. 9. (a) PCA basis (b) PCA reduction to 1D
(a) The concept of PCA. Solid lines: the original basis; dashed lines: the PCA basis. The dots are selected at regularly spaced locations on a straight line rotated at 30degrees, and then perturbed by isotropic 2D Gaussian noise. (b) The projection (1D reconstruction) of the data using only the first principal component.

There are many different algorithms available for Gesture Recognition, such as principal Component Analysis, Linear Discriminate Analysis, Independent Component Analysis, Neural Networks and Genetic Algorithms.

Principal Component Analysis (PCA)[11] is used in this application to compare the acquired gesture with the gestures available in the database and recognize the intended gesture.

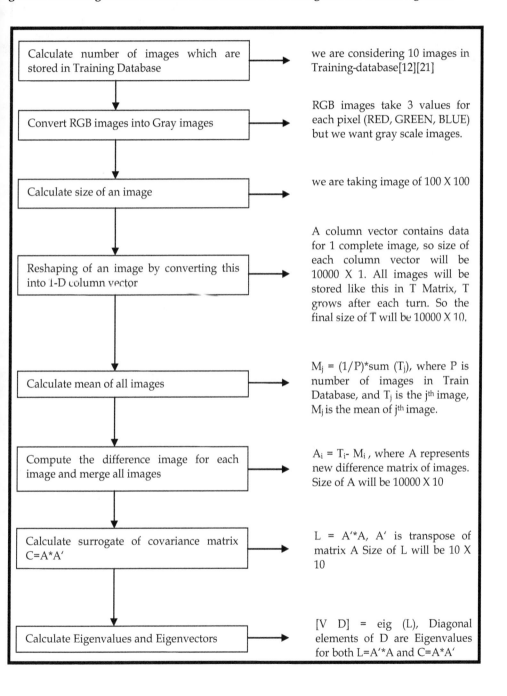

| | |
|---|---|
| Calculate number of images which are stored in Training Database | we are considering 10 images in Training-database[12][21] |
| Convert RGB images into Gray images | RGB images take 3 values for each pixel (RED, GREEN, BLUE) but we want gray scale images. |
| Calculate size of an image | we are taking image of 100 X 100 |
| Reshaping of an image by converting this into 1-D column vector | A column vector contains data for 1 complete image, so size of each column vector will be 10000 X 1. All images will be stored like this in T Matrix, T grows after each turn. So the final size of T will be 10000 X 10. |
| Calculate mean of all images | $M_j = (1/P)*$sum $(T_j)$, where P is number of images in Train Database, and $T_j$ is the $j^{th}$ image, $M_j$ is the mean of $j^{th}$ image. |
| Compute the difference image for each image and merge all images | $A_i = T_i - M_i$ , where A represents new difference matrix of images. Size of A will be 10000 X 10 |
| Calculate surrogate of covariance matrix C=A*A′ | L = A′*A, A′ is transpose of matrix A Size of L will be 10 X 10 |
| Calculate Eigenvalues and Eigenvectors | [V D] = eig (L), Diagonal elements of D are Eigenvalues for both L=A′*A and C=A*A′ |

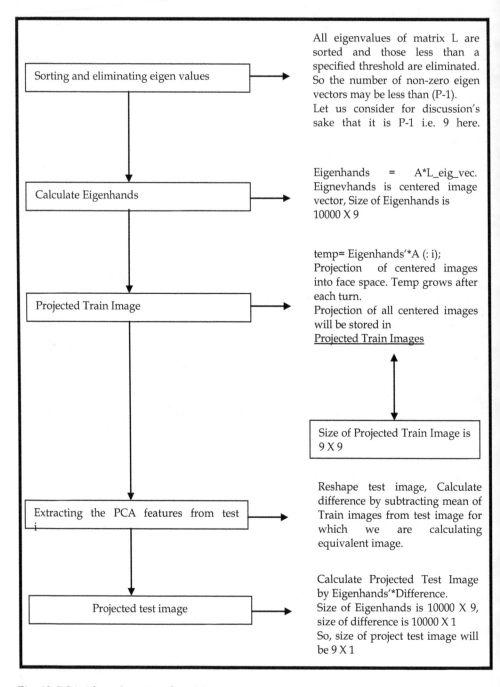

Fig. 10. PCA Algorithm steps for 10 Images

The concept of PCA is illustrated in Figure 9; The graph shows data that originally has two components along the $x_1$ and $x_2$ directions. This is now resolved along the $\Phi_1$ and $\Phi_2$ directions. $\Phi_1$ corresponds to the direction of maximum variance and is chosen as the first principal component. In the 2D case currently being considered, the second principal component is then determined uniquely by orthogonality constraints; in a higher-dimensional space the selection process would continue, guided by the variances of the projections.

A comprehensive explanation of the above scheme for gesture recognition[19] within a database of ten images is given in figure 10. For simplicity of calculations and ease of understanding, each figure is taken to be of size of 100X100 pixels. This algorithms was tested on a PC platform.

## 2.4 Generation of control Instructions for a robot

Different functions corresponding to each meaningful hand gesture are written and stored in database for controlling the robotic Arm. For implementation of robotic arm control, PUMA robotic model has been chosen. Since this model was only available at the time of interface, hand gestures are used to control the various joints of the arm. However, these gestures can be used to control the robotic hand also. Altogether 10 different hand gestures are related to 10 different movements of robotic arm. Movement commands are written as a function in robot specific language. Whenever a gesture is matched with a meaningful gesture from the database, the instruction set corresponding to that gesture is identified and passed to robot for execution. In this way the robotic system can be controlled by hand gesture using live camera. Figure 11 shows movements of robotic hand corresponding to four different specific hand gestures.

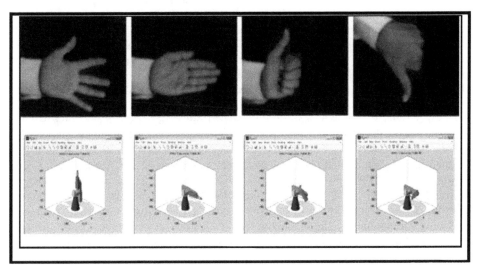

Fig. 11. Movements of PUMA 762 corresponding to some specific hand gestures

However, there are a total of 10 actionable gestures, and one background image for resetting purpose that have been identified as shown in fig. 12.

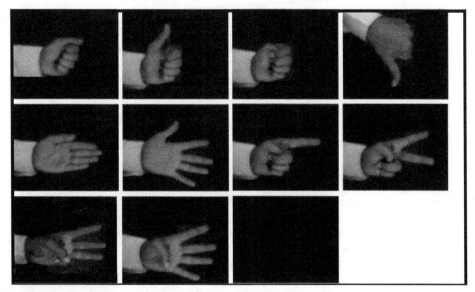

Fig. 12. Set of specified gestures used for robot control

## 3. Implementation on FPGA processor

We have implemented the algorithm on an FPGA, using a Database of 6 images. Each image size is 60X80 pixels. The reduction in number of images, image size and its resolution are due to FPGA memory constraints. The MATLAB code produces 6x6 projection vectors which are stored in the RAM of the FPGA using a JTAG cable [27]. After formation of the database, the system captures an image and projection vector of 6x1 is produced and that is given to the hardware for further processing. The hardware is responsible for calculating the Euclidian Distance [ED] with a dedicated block which we designate as ED block [34]. Then the minimum distance [MD] is found using a dedicated block named MD block. And the index corresponding to the minimum distance is returned to MATLAB to display the matched image from the Database [33]. The whole process is shown in fig. 13.

The main drawback of this implementation is that only part of the algorithm resides on FPGA and rest of the algorthim still runs on PC. Therefore it is not standalone system and requires interface with the PC[13]. FPGA implementation of PCA algorithm is a great challenge  due to the limited resources available in commercial FPGA kits. But it is quite important to have a complete system on a FPGA chip [40]. As far as the technology is concerned, FPGA is most suited for implementation of real time image processing algorithm. It can be readily designed with custom parallel digital circuitry tailored for performing various imaging tasks making them well-suited for high speed real time vision processing applications[35].

## 4. Conclusion

Controlling a robot arm, in real time,  through the hand gestures is a novel approach. The technique proposed here was tested under proper lighting conditions that we created in our laboratory. A gesture database consisting of binary images of size 60 X 80 pixels is pre-stored, so it takes less time and memory space during pattern recognition. Due to the use of cropped image of gesture, our database becomes more effective as one image is sufficient for one type of gesture presentation. So, neither we need to store more than one image for same gesture at different positions of image, nor have we to worry about positions of hand in front of camera. Experimental results show that the system can detect hand gestures with an accuracy of 95 % when it is kept still in front of the camera for 1 second.

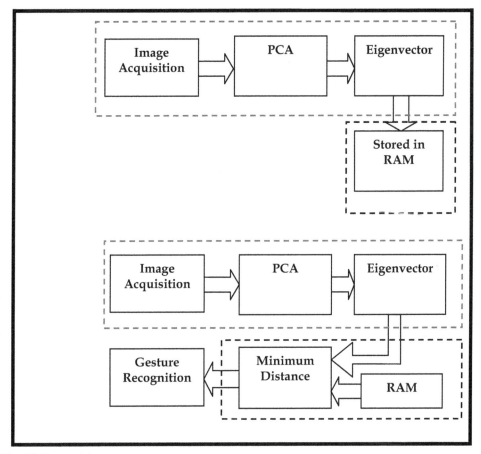

Fig. 13. Recognition process

## 5. Acknowledgment

This research is being carried out at Central Electronics Engineering Research Institute (CEERI), Pilani, India as part of a project "Supra Institutional Projects on Technology Development for Smart Systems". Authors would like to thank Director, CEERI for his active encouragement and support.

# 6. References

[1] Becerra V.M., Cage, C.N.J., Harwin, W.S., Sharkey, P.M. Hardware retrofit and computed torque control of Puma 560 robot updating an industrial manipulator. IEEE Control Systems Magazine, 24(5): 78-82. 2004.

[2] L. Brthes, P. Menezes, F. Lerasle, and J. Hayet. Face tracking and hand gesture recognition for human robot interaction. In International Conference on Robotics and Automation, pp. 1901–1906, 2004. New Orleans, Louisiana.

[3] Chan Wah Ng, Surendra Ranganath, Real-time gesture recognition system and application. Image and Vision computing (20): 993-1007, 2002.

[4] Corke, P.I. Operational Details of the Unimation Puma Servo System. Report, CSIRO Division of Manufacturing Technology. Australia, 1993.

[5] Crow, F. Summed-area tables for texture mapping. SIGGRAPH '84: Proceedings of the 11th annual conference on Computer graphics and interactive techniques. pp. 207-212.

[6] J. Davis, M. Shah. Recognizing hand gestures. In Proceedings of European Conference on Computer Vision, ECCV: 331-340, 1994. Stockholm, Sweden.

[7] Daggu Venkateshwar Rao, Shruti Patil, Naveen Anne Babu and V Muthukumar, Implementation and Evaluation of Image Processing Algorithms on Reconfigurable Architecture using C-based Hardware Descriptive Languages, International Journal of Theoretical and Applied Computer Sciences, Volume 1 Number 1 pp. 9–34, 2006.

[8] Dastur, J.; Khawaja, A.; Robotic Arm Actuation with 7 DOF Using Haar Classifier Gesture Recognition, Second International Conference on Computer Engineering and Applications (ICCEA), vol.1, no., pp.27-29, 19-21 March 2010.

[9] K G Derpains, A review of Vision-based Hand Gestures, Internal Report, Department of Computer Science. York University, February 2004.

[10] Didier Coquin, Eric Benoit, Hideyuki Sawada, and Bogdan Ionescu, Gestures Recognition Based on the Fusion of Hand Positioning and Arm Gestures, Journal of Robotics and Mechatronics Vol.18 No.6, 2006. pp. 751-759.

[11] Fang Zhong, David W. Capson, Derek C. Schuurman, Parallel Architecture for PCA Image Feature Detection using FPGA 978-1-4244-1643-1/08/2008 IEEE.

[12] Freeman W., Computer vision for television and games. In Recognition, Analysis, and Tracking of Faces and Gestures in Real-Time Systems, page 118, 1999. Copyright© MEITCA.

[13] Haessig D.,Hwang J., Gallagher S., Uhm M., Case-Study of a Xilinx System Generator Design Flow For Rapid Development of SDR Waveforms, Proceeding of the SDR 05 Technical Conference and Product Exposition. Copyright © 2005 SDR Forum.

[14] Jae-Ho Shin, Jong-Shill Lee, Se-Kee Kil, Dong-Fan Shen, Je-Goon Ryu, Eung-Hyuk Lee, Hong-Ki Min, Seung-Hong Hong, Hand Region Extraction and Gesture Recognition using entropy analysis, IJCSNS International Journal of Computer Science and Network Security, VOL.6 No.2A, 2006 pp. 216-222.

[15] Jagdish Raheja, Radhey Shyam and Umesh Kumar, Hand Gesture Capture and Recognition Technique for Real-time Video Stream, In The 13th IASTED International Conference on Artificial Intelligence and Soft Computing (ASC 2009), September 7 - 9, 2009 Palma de Mallorca, Spain.

[16] Katupitiya, J., Radajewski, R., Sanderson, J., Tordon, M. Implementation of PC Based Controller for a PUMA Robot. Proc. 4th IEEE Conf. on Mechatronics and Machine Vision in Practice. Australia, p.14-19. [doi:10.1109/MMVIP.1997.625229], 1997.

[17] R. Kjeldsen, J. Kinder. Visual hand gesture recognition for window system control, in International Workshop on Automatic Face and Gesture Recognition (IWAFGR), Zurich: pp. 184-188, 1995.

[18] Mohammad, Y.; Nishida, T.; Okada, S.; Unsupervised simultaneous learning of gestures, actions and their associations for Human-Robot Interaction, Intelligent Robots and Systems, 2009. IROS 2009. IEEE/RSJ International Conference on, vol., no., pp.2537-2544, 10-15 Oct. 2009.

[19] Moezzi and R. Jain, "ROBOGEST: Telepresence Using Hand Gestures". Technical report VCL-94-104, University of California, San Diego, 1994.

[20] M. Moore, A DSP-based real time image processing system. In the Proceedings of the 6th International conference on signal processing applications and technology, Boston MA, August 1995.

[21] Nolker, C.; Ritter, H.; Parametrized SOMs for hand posture reconstruction, Neural Networks, 2000. IJCNN 2000, Proceedings of the IEEE-INNS-ENNS International Joint Conference on , vol.4, no., pp.139-144 vol.4, 2000

[22] Nolker C., Ritter H., Visual Recognition of Continuous Hand Postures, IEEE Transactions on neural networks, Vol 13, No. 4, July 2002, pp. 983-994.

[23] K. Oka, Y. Sato and H. Koike. Real-Time Fingertip Tracking and Gesture Recognition. IEEE Computer Graphics and Applications: 64-71, 2002.

[24] Ozer, I. B., Lu, T., Wolf, W. Design of a Real Time Gesture Recognition System: High Performance through algorithms and software. IEEE Signal Processing Magazine, pp. 57-64, 2005.

[25] V.I. Pavlovic, R. Sharma, T.S. Huang. Visual interpretation of hand gestures for human-computer interaction, A Review, IEEE Transactions on Pattern Analysis and Machine Intelligence 19(7): 677-695, 1997.

[26] Quek, F. K. H., Mysliwiec, T., Zhao, M. (1995). Finger Mouse: A Free hand Pointing Computer Interface. Proceedings of International Workshop on Automatic Face and Gesture Recognition, pp.372-377, Zurich, Switzerland.

[27] A.K. Ratha, N.K. Jain, FPGA-based computing in computer vision, Computer Architecture for Machine Perception, CAMP '97. Proceedings Fourth IEEE International Workshop on, pp. 128–137, 20-22 Oct 1997.

[28] Ramamoorthy, N. Vaswani, S. Chaudhury, S. Banerjee, Recognition of Dynamic hand gestures, Pattern Recognition 36: 2069-2081, 2003.

[29] Sawah A.E., and et al., a framework for 3D hand tracking and gesture recognition using elements of genetic programming, 4th Canadian conference on Computer and robot vision, Montreal, Canada, 28-30 May, 2007, pp. 495-502.

[30] Sergios Theodoridis, Konstantinos Koutroumbas, Pattern Recognition, Elsevier Publication, Second Edition, 2003.

[31] Seyed Eghbal Ghobadi, Omar Edmond Loepprich, Farid Ahmadov Jens Bernshausen, Real Time Hand Based Robot Control Using Multimodal Images, IAENG International Journal of Computer Science, 35:4, IJCS_35_4_08, Nov 2008.

[32] Shin M. C., Tsap L. V., and Goldgof D. B., Gesture recognition using bezier curves for visualization navigation from registered 3-d data, Pattern Recognition, Vol. 37, Issue 5, May 2004, pp.1011–1024.

[33] N. Shirazi and J. Ballagh, "Put Hardware in the Loop with Xilinx System Generator for DSP", Xcell Journal online, May 2003,
    hitp://xvww.xilinix.coinjpoblication s/xcellonlince/xcell 47/xc svsen47.htm

[34] Stephen D.Brown, R.J. Francis, J.Rose, Z.G.Vranesic, Field  Programmable Gate Arrays, 1992.

[35] Starner T., Pentland. A., Real-time American Sign Language recognition from video using hidden markov models. In SCV95, pages 265–270, Ames Street, Cambridge, 1995.

[36] Triesch J., Malsburg. C.V.D., Robotic gesture recognition. In Gesture Workshop, pages 233–244, Nara, Japan, 1997.

[37] Tomita, A., Ishii. R. J. Hand shape extraction from a sequence of digitized  grayscale images. Proceedings of tweenth International Conference on Industrial Electronics, Control and Instrumentation, IECON '94, vol.3, pp.1925–1930, Bologna, Italy, 1994.

[38] Verma R., Dev A., Vision based Hand Gesture Recognition Using finite State Machines and Fuzzy Logic, International Conference on Ultra-Modern Telecommunications & Workshops, 12-14 Oct, 2009, pp. 1-6, in St. Petersburg, Russia.

[39] Walla Walla University, robotic lab PUMA 762.
    http://www.mathworks.com/matlabcentral/fileexchange/14932

[40] Xilinx Inc., System Generator for DSP, Users Guide
    http://wsssw.xilinx.con1i/products/softwa-e/svstzen/app docs/userLguide.htm

[41] Ying Wu, Thomas S Huang, Vision based Gesture. Recognition: A Review, Lecture Notes In Computer Science; Vol. 1739, Proceedings of the International Gesture Workshop on Gesture-Based Communication in Human-Computer Interaction, 1999.

# Part 2

# Programming and Algorithms

# 4

# Robotic Software Systems: From Code-Driven to Model-Driven Software Development

Christian Schlegel, Andreas Steck and Alex Lotz
*Computer Science Department, University of Applied Sciences Ulm*
*Germany*

## 1. Introduction

Advances in robotics and cognitive sciences have stimulated expectations for emergence of new generations of robotic devices that interact and cooperate with people in ordinary human environments (robot companion, elder care, home health care), that seamlessly integrate themselves into complex environments (domestic, outdoor, public spaces), that fit into different levels of system hierarchies (human-robot co-working, hyper-flexible production cells, cognitive factory), that can fulfill different tasks (multi-purpose systems) and that are able to adapt themselves to different situations and changing conditions (dynamic environments, varying availability and accessibility of internal and external resources, coordination and collaboration with other agents).

Unfortunately, so far, steady improvements in specific robot abilities and robot hardware have not been matched by corresponding robot performance in real-world environments. On the one hand, simple robotic devices for tasks such as cleaning floors and cutting the grass have met with growing commercial success. Robustness and single purpose design is the key quality factor of these simple systems. At the same time, more sophisticated robotic devices such as *Care-O-Bot 3* (Reiser et al., 2009) and *PR2* (Willow Garage, 2011) have not yet met commercial success. Hardware and software complexity is their distinguishing factor.

Advanced robotic systems are systems of systems and their complexity is tremendous. Complex means they are built by integrating an increasingly larger body of heterogeneous (robotics, cognitive, computational, algorithmic) resources. The need for these resources arises from the overwhelming number of different situations an advanced robot is faced with during execution of multitude tasks. Despite the expended effort, even sophisticated systems are still not able to perform at an expected and appropriate level of overall quality of service in complex scenarios in real-world environments. By quality of service we mean the set of system level non-functional properties that a robotic system should exhibit to appropriately operate in an open-ended environment, such as robustness to exceptional situations, performance despite of limited resources and aliveness for long periods of time.

Since vital functions of advanced robotic systems are provided by software and software dominance is still growing, the above challenges of system complexity are closely related to the need of mastering software complexity. Mastering software complexity becomes pivotal towards exploiting the capabilities of advanced robotic components and algorithms. Tailoring modern approaches of software engineering to the needs of robotics is seen as decisive towards significant progress in system integration for advanced robotic systems.

## 2. Software engineering in robotics

Complex systems are rarely built from scratch but their design is typically partitioned according to the variety of technological concerns. In robotics, these are among others mechanics, sensors and actuators, control and algorithms, computational infrastructure and software systems. In general, successful engineering of complex systems heavily relies on the *divide and conquer* principle in order to reduce complexity. Successful markets typically come up with precise role assignments for participants and stakeholders ranging from component developers over system integrators and experts of an application domain to business consultants and end-users.

Sensors, actuators, computers and mechanical parts are readily available as commercial off-the-shelf black-box components with precisely specified characteristics. They can be re-used in different systems and they are provided by various dedicated suppliers. In contrast, most robotics software systems are still based on proprietarily designed software architectures. Very often, robotics software is tightly bound to specific robot hardware, processing platforms, or communication infrastructures. In addition, assumptions and constraints about tasks, operational environments, and robotic hardware are hidden and hard-coded in the software implementation.

Software for robotics is typically embedded, concurrent, real-time, distributed, data-intensive and must meet specific requirements, such as safety, reliability and fault-tolerance. From this point of view, software requirements of advanced robots are similar to those of software systems in other domains, such as avionics, automotive, factory automation, telecommunication and even large scale information systems. In these domains, modern software engineering principles are rigorously applied to separate roles and responsibilities in order to cope with the overall system complexity.

In robotics, tremendous code-bases (libraries, middleware, etc.) coexist without being interoperable and each tool has attributes that favors its use. Although one would like to reuse existing and matured software building blocks in order to reduce development time and costs, increase robustness and take advantage from specialized and second source suppliers, up to now this is not possible. Typically, experts for application domains need to become experts for robotics software to make use of robotics technology in their domain. So far, robotics software systems even do not enforce separation of roles for component developers and system integrators.

The current situation in software for robotics is caused by the lack of separation of concerns. In consequence, role assignments for robotics software are not possible, there is nothing like a software component market for robotic systems, there is no separation between component developers and system integrators and even no separation between experts in robotics and experts in application domains. This is seen as a major and serious obstacle towards developing a market of advanced robotic systems (for example, all kinds of cognitive robots, companion systems, service robots).

The current situation in software for robotics can be compared with the early times of the *World Wide Web (WWW)* where one had to be a computer engineer to setup web pages. The WWW turned into a universal medium only since the availability of tools which have made it accessible and which support separation of concerns: domain experts like journalists can now easily provide content without bothering with technical details and there is a variety of specialized, competing and interoperable tools available provided by computer engineers, designers and others. These can be used to provide and access any kind of content and to support any kind of application domain.

Based on these observations, we assume that the next big step in advanced robotic systems towards mastering their complexity and their overall integration into any kind of environment and systems depends on separation of concerns. Since software plays a pivotal role in advanced robotic systems, we illustrate how to tailor a service-oriented component-based software approach to robotics, how to support it by a model-driven approach and according tools and how this allows separation of concerns which so far is not yet addressed appropriately in robotics software systems.
*Experienced software engineers* should get insights into the specifics of robotics and should better understand what is in the robotics community needed and expected from the software engineering community. *Experienced roboticists* should get detailed insights into how model-driven software development *(MDSD)* and its design abstraction is an approach towards system-level complexity handling and towards decoupling of robotics knowledge from implementational technologies. *Practitioners* should get insights into how separation of concerns in robotics is supported by a service-oriented component-based software approach and that according tools are already matured enough to make life easier for developers of robotics software and system integrators. *Experts in application domains* and *business consultants* should gain insights into maturity levels of robotic software systems and according approaches under a short-term, medium-term and long-term perspective. *Students* should understand how design abstraction as recurrent principle of computer science applied to software systems results in *MDSD*, how *MDSD* can be applied to robotics, how it provides a perspective to overcome the vicious circle of robotics software starting from scratch again and again and how software engineering and robotics can cross-fertilize each other.

## 2.1 Separation of concerns

Separation of concerns is one of the most fundamental principles in software engineering (Chris, 1989; Dijkstra, 1976; Parnas, 1972). It states that a given problem involves different kinds of concerns, promotes their identification and separation in order to solve them separately without requiring detailed knowledge of the other parts, and finally combining them into one result. It is a general problem solving strategy which breaks the problem complexity into loosely-coupled subproblems. The solutions to the subproblems can be composed relatively easily to yield a solution to the original problem (Mili et al., 2004). This allows to cope with complexity and thereby achieving the required engineering quality factors such as robustness, adaptability, maintainability, and reusability.

Despite a common agreement on the necessity of the application of the separation of concerns principle, there is not a well-established understanding of the notion of concern. Indeed, *concern* can be thought of as a unit of modularity (Blogspot, 2008). Progress towards separation of concerns is typically achieved through modularity of programming and encapsulation (or *transparency* of operation), with the help of information hiding. Advanced uses of this principle allow for simultaneous decomposition according to multiple kinds of (overlapping and interacting) concerns (Tarr et al., 2000).

In practice, the principle of separation of concerns should drive the identification of the right decomposition or modularization of a problem. Obviously, there are both: (i) generic and domain-independent patterns of how to decompose and modularize certain problems in a suitable way as well as (ii) patterns driven by domain-specific best practices and use-cases.

In most engineering approaches as well as in robotics, at least the following are dominant dimensions of concerns which should be kept apart (Björkelund et al., 2011; Radestock & Eisenbach, 1996):

**Computation** provides the functionality of an entity and can be implemented in different ways (software and/or hardware). Computation activities require communication to access required data and to provide computed results to other entities.

**Communication** exchanges data between entities (ranging from hardware devices to interfaces for real-world access over software entities to user interfaces etc.).

**Configuration** comprises the binding of configurable parameters of individual entities. It also comprises the binding of configurable parameters at a system level like, for example, connections between entities.

**Coordination** is about when is something being done. It determines how the activities of all entities in a system should work together. It relates to orchestration and resource management.

According to (Björkelund et al., 2011), this is in line with results published in (Delamer & Lastra, 2007; Gelernter & Carriero, 1992; Lastra & Delamer, 2006) although variations exist which split *configuration* (into *connection* and *configuration*) or treat *configuration* and *coordination* in the same way (Andrade et al., 2002; Bruyninckx, 2011).

It is important to recognize that there are cross-cutting concerns like *quality of service (QoS)* that have instantiations within the above dimensions of concerns. Facets of *QoS for computation* can manifest with respect to time (best effort computation, hard real-time computation) or anytime algorithms (explicated relationship between assigned computing resources and achieved quality of result). Facets of *QoS for communication* are, for example, response times, latencies and bandwidth.

It is also important to recognize that various concerns need to be addressed at different stages of the lifecycle of a system and by different stakeholders. For example, *configuration* is part of the *design phase* (a component developer provides dedicated configurable parameters, a system integrator binds some of them for deployment) *and* of the *runtime phase* (the task coordination mechanism of a robot modifies parameter settings and changes the connections between entities according to the current situation and task to fulfill).

It is perfectly safe to say that robotics should take advantage from insights and successful approaches for complexity handling readily available in other but similar domains like, for example, automotive and avionics industry or embedded systems in general. Instead, robotics often reinvents the wheel instead of exploiting cross-fertilization between robotics and communities like software engineering and middleware systems. The interesting question is whether there are differences in robotics compared to other domains which hinder roboticists from jumping onto already existing and approved solutions. One should also examine whether or not these solutions are tailorable to robotics needs.

### 2.2 Specifics in robotics

The difference of robotics compared to other domains like automotive and avionics is neither the huge variety of different sensors and actuators nor the number of different disciplines being involved nor the diversity of hardware-platforms and software-platforms. In many domains, developers need to deal with heterogeneous hardware devices and are obliged to deploy their software on computers which are often constrained in terms of memory and computational power.

We are convinced that differences of robotics compared to other domains originate from the need of a robot to cope with open-ended environments while having only limited resources at its disposal.

Limited resources require decisions: when to assign which resources to what activity taking into account perceived situation, current context and tasks to be fulfilled. Finding adequate solutions for this major challenge of engineering robotic systems is difficult for two reasons:

- the *problem space* is huge: as uncertainty of the environment and the number and type of resources available to a robot increase, the definition of the best matching between current situation and correct robot resource exploitation becomes an overwhelming endeavour even for the most skilled robot engineer,

- the *solution space* is huge: in order to enhance overall quality of service like robustness of complex robotic systems in real-world environments, robotic system engineers should master highly heterogeneous technologies, need to integrate them in a consistent and effective way and need to adequately exploit the huge variety of robotic-specific resources.

In consequence, it is impossible to statically assign resources in advance in such a way that all potential situations arising at runtime are properly covered. Due to open-ended real-world environments, there will always be a deviation between *design-time optimality* and *runtime optimality* with respect to resource assignments. Therefore, there is a need for dynamic resource assignments at runtime which arises from the enormous sizes of the problem space and the solution space.

For example, a robot designer cannot foresee how crowded an elevator will be. Thus, a robot will need to decide by its own and at runtime whether it is possible and convenient to exploit the elevator resource. The robot has to trade the risk of hitting an elevator's user with the risk of arriving late at the next destination. To match the level of safety committed at design-time, the runtime trade-off has to come up with parameters for speed and safety margins whose risk is within the design-time committed boundaries while still implementing the intent to enter the elevator.

### 2.2.1 Model-centric robotic systems

The above example illustrates why we have to think of engineering advanced robotic systems differently compared to other complex systems. A complex robotic system cannot be treated as design-time finalizable system. At runtime, system configurations need to be changeable according to current situation and context including prioritized assignments of resources to activities, (de)activations of components as well as changes to the wiring between components. At runtime, the robot has to analyze and to decide for the most appropriate configuration. For example, if the current processor load does not allow to run the navigation component at the highest level of quality, the component should be configured to a lower level of navigation quality. A reasonable option to prepare a component to cope with reduced resource assignments might be to reduce the maximum velocity of the robot in order to still guarantee the same level of navigation safety.

In consequence, we need to support design-time reasoning (at least by the system engineer) as well as runtime reasoning (by the robot itself) about both, the problem space and the solution space. This can be achieved by raising the level of abstraction at which relevant properties and characteristics of a robotics system are expressed. As for every engineering endeavour, this means to rely on the power of models and asks for an overall different design approach as illustrated in figure 1:

- The solution space can be managed by providing advanced design tools for robot software development to design reconfigurable and adaptive robotic systems. Different stakeholders involved in the development of a robotic system need the ability to formally

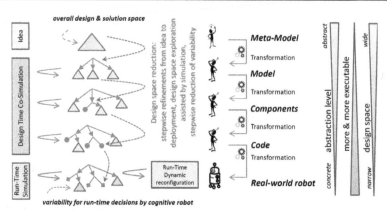

Fig. 1. Novel workflow bridging design-time and runtime model-usage: at design-time variation points are purposefully left open and allow for runtime decisions (Schlegel et al., 2010).

model and relate different views relevant to robotic system design. A major issue is the support of separation of concerns taking into account the specific needs of robotics.

- The problem space can be mastered by giving the robot the ability to reconfigure its internal structure and to adapt the way its resources are exploited according to its understanding of the current situation.

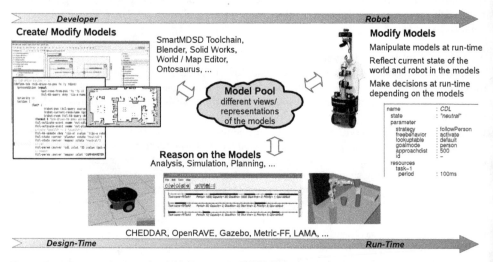

Fig. 2. Separation of concerns and design abstraction: models created at design-time are used and manipulated at runtime by the robot (Steck & Schlegel, 2011).

We coin the term *model-centric robotic systems* (Steck & Schlegel, 2011) for the new approach of using models to cover and support the whole life-cycle of robotic systems. Such a model-centric view puts models into focus and bridges design-time and runtime model-usage.

During the whole lifecycle, models are refined and enriched step-by-step until finally they become executable. Models comprise variation points which support alternative solutions. Some variation points are purposefully left open at design time and even can be bound earliest at runtime after a specific context and situation dependent information is available. In consequence, models need to be interpretable not only by a human designer but also by a computer program. At design-time, software tools should understand the models and support designers in their transformations. At runtime, adaptation algorithms should exploit the models to automatically reconfigure the control system according to the operational context (see figure 2).

The need to explicitly support the design for runtime adaptability adds robotic-specific requirements on software structures and software engineering processes, gives guidance on how to separate concerns in robotics and allows to understand where the robotics domain needs extended solutions compared to other and at first glance similar domains.

### 2.2.2 User roles and requirements

Another strong influence on robotic software systems besides technical challenges comes from the involved individuals and their needs. We can distinguish several user roles that all put a different focus on complexity management, on separation of concerns and on software engineering in robotics:

**End users** operate applications based on the provided user interface. They focus on the functionality of readily provided systems. They do not care on how the application has been built and mainly expect reliable operation, easy usage and reasonable value for money.

**Application builders / system integrators** assemble applications out of approved, standardized and reusable off-the-shelf components. Any non trivial robotic application requires the orchestration of several components such as computer vision, sensor fusion, human machine interaction, object recognition, manipulation, localization and mapping, control of multiple hardware devices, etc. Once these parts work together, we call it a *system*. This part of the development process is called, therefore, *system integration*. Components can be provided by different vendors. Application builders and system integrators consider components as black boxes and depend on precise specifications and explications of all relevant properties for smooth composition, resource assignments and mappings to target platforms. Components are customized during system level composition by adjusting parameters or filling in application dependent parts at so-called *hot spots* via plug-in interfaces. Application builders expect support for system-level engineering.

**Component builders** focus on the specification and implementation of a single component. They want to focus on algorithms and component functionality without being restricted too much with respect to component internals. They don't want to bother with integration issues and expect a framework to support their implementation efforts such that the resulting component is conformant to a system-level black box view.

**Framework builders / tool providers** prepare and provide tools that allow the different users to focus on their role. They implement the software frameworks and the domain-specific add-ons on top of state-of-the-art and standard software systems (like middleware systems), use latest software technology and make these available to the benefit of robotics.

**The robotics community** provides domain-specific concepts, best practices and design patterns of robotics. These are independent of any software technology and implementational technology. They form the *body of knowledge of robotics* and provide the domain-specific ground for the above roles.

The essence of the work of the *component builder* is to design reusable components which can seamlessly be integrated into multiple systems and different hardware platforms. A component is considered as a black box. The developer can achieve this abstraction only if he is strictly limited in his knowledge and assumptions about what happens outside his component and what happens inside other components.

On the other hand, the methodology and the purpose of the *system integrator* is opposite: he knows exactly the application of the software system, the platform where it will be deployed and its constraints. For this reason, he is able to take the right decision about the kind of components to be used, how to connect them together and how to configure their parameters and the quality of service of each of them to orchestrate their behavior. The work of the system integrator is rarely reusable by others, because it is intrinsically related to a specific hardware platform and a well-defined and sometimes unique use-case. We don't want the system integrator to modify a component or to understand the internal structure and implementation of the components he assembles.

### 2.2.3 Separation of roles from an industrial perspective

This distinction between the development of single components and system integration is important (figure 3). So far, reuse in robotics software is mainly possible at the level of libraries and/or complete frameworks which require system integrators to be component developers and vice versa. A formal separation between *component building* and *system integration* introduces another and intermediate level of abstraction for reuse which will make it possible to

- create commercial off-the-shelf (COTS) robotic software: when components become independent of any specific robot application, it becomes possible to integrate them quickly into different robotic systems. This abstraction allows the component developer to sell its robotic software component to a system integrator;

- overcome the need for the system integrator to be also an expert of robotic algorithms and software development. We want companies devoted to system integration (often SMEs) to take care of the Business-to-Client part of the value chain, but this will be possible only when their work will become less challenging;

- establish dedicated system integrators (specific to industrial branches and application domains) apart from experts for robotic components (like navigation, localization, object recognition, speech interaction, etc.);

- provide plug-and-play robotic hardware: so far the effort of the integration of the hardware into the platform was undertaken by the system integrator. If manufacturers start providing ready-to-use drivers which work seamlessly in a component-driven environment, robotic applications can be deployed faster and become cheaper.

This separation of roles will eventually have a positive impact in robotics: it will potentially allow the creation of a robotics industry, that is an ecosystem of small, medium and large enterprises which can profitably and symbiotically coexist to provide business-to-business

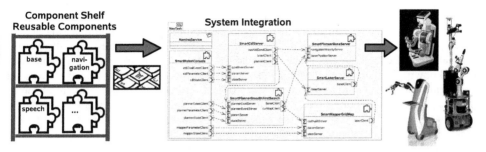

Fig. 3. Building robotic systems out of readily-available and reusable software components: separation of the roles of component development and system integration.

and business-to-client products such as hardware parts, software applications and complete robots with a specific application and deliver a valuable product to the customer.

To understand better what a *robotics industry* means, we draw an analogy to the personal computer industry. Apart from very few exceptions, we can identify several companies involved in the manufacturing of a single and very specific part of the final product: single hardware components (memories, hard drives, CPU, mother boards, screens, power supplies, graphic cards, etc.), operating systems (Windows, commercial Linux distributions), software applications (CAD, word processing, video games, etc.) and system integrators which provide ready-to-use platforms to the end user.

### 2.3 Service-oriented software components to master system complexity

Software engineering provides three major approaches that help to address the above challenges, that is *component-based software engineering (CBSE)*, *service-oriented architectures (SOA)* and *model-driven software development (MDSD)*.

CBSE separates the component development process from the system development process and aims at component reusability. MDSD separates domain knowledge (formally specified by domain experts) from how it is being implemented (defined by software experts using model transformations). SOA is about the right level of granularity for offering functionality and strictly separates service providers and consumers.

### 2.3.1 Component-based software engineering

CBSE (Heineman & Councill, 2001) is an approach that has arisen in the software engineering community in the last decade. It shifts the emphasis in system-building from traditional requirements analysis, system design and implementation to composing software systems from a mixture of reusable off-the-shelf and custom-built components. A compact and widely accepted definition of a software component is the following one:

"A **software component** is a unit of composition with contractually specified interfaces and explicit context dependencies only. A software component can be developed independently and is subject to composition by third parties." (Szyperski, 2002).

*Software components* explicitly consider reusable pieces of software including notions of independence and late composition. *CBSE* promises the benefits of increased reuse, reduced production cost, and shorter time to market. In order to realize these benefits, it is vital to have components that are easy to reuse and composition mechanisms that can be applied systematically. Composition can take place during different stages of the lifecycle of components that is during the design phase (design and implementation), the deployment phase (system integration) and even the runtime phase (dynamic wiring of data flow according to situation and context). *CBSE* is based on the explication of all relevant information of a component to make it usable by other software elements whose authors are not known. The key properties of *encapsulation* and *composability* result in the following seven criteria that make a good component: "(i) may be used by other software elements (clients), (ii) may be used by clients without the intervention of the component's developers, (iii) includes a specification of all dependencies (hardware and software platform, versions, other components), (iv) includes a precise specification of the functionalities it offers, (v) is usable on the sole basis of that specification, (vi) is composable with other components, (vii) can be integrated into a system quickly and smoothly" (Meyer, 2000).

### 2.3.2 Service-oriented architectures

Another generally accepted view of a software component is that it is a software unit with *provided services* and *required services*. In component models, where components are architectural units, *services* are represented as *ports* (Lau & Wang, 2007). This view puts the focus on the question of a proper level of abstraction of offered functionalities. Services "combine information and behavior, hide the internal workings from outside intrusion and present a relatively simple interface to the rest of the program" (Sprott & Wilkes, 2004). The (CBDI Forum, 2011) recommends to define *service-oriented architectures (SOA)* as follows:

**SOA** are "the policies, practices, frameworks that enable application functionality to be provided and consumed as sets of services published at a granularity relevant to the service consumer. Services can be invoked, published and discovered, and are abstracted away from the implementation using a single, standards-based form of interface" (Sprott & Wilkes, 2004).

*Service* is the key to communication between providers and consumers and key properties of good service design are summarized as in table 1. *SOA* is all about style (policy, practice, frameworks) which makes process matters an essential consideration. A *SOA* has to ensure that services don't get reduced to the status of interfaces, rather they have an identity of their own. With *SOA*, it is critical to implement processes that ensure that there are at least two different and separate processes - for providers and consumers (Sprott & Wilkes, 2004).

| reusable | use of service, not reuse by copying of code/implementation |
|---|---|
| abstracted | service is abstracted from the implementation |
| published | precise, published specification functionality of service interface, not implementation |
| formal | formal contract between endpoints places obligations on provider and consumer |
| relevant | functionality is presented at a granularity recognized by the user as a meaningful service |

Table 1. Principles of good service design enabled by characteristics of *SOA* as formulated in (Sprott & Wilkes, 2004).

### 2.3.3 Model-driven software development

*MDSD* is a technology that introduces significant efficiencies and rigor to the theory and practice of software development. It provides a design abstraction as illustrated in figure

4. Abstractions are provided by models (Beydeda et al., 2005). Abstraction is a core principle of software engineering.

"A **model** is a simplified representation of a system intended to enhance our ability to understand, predict and possibly control the behavior of the system" (Neelamkavil, 1987).

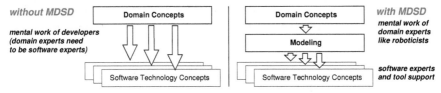

Fig. 4. Design abstraction of model-driven software development.

In *MDSD*, models are used for many purposes, including reasoning about problem and solution domains and documenting the stages of the software lifecycle; the result is improved software quality, improved time-to-value and reduced costs (IBM, 2006).

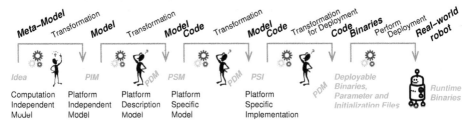

Fig. 5. Model-driven software development at a glance.

The standard workflow of a model-driven software development process is illustrated in figure 5. This workflow is supported by tools like the *Eclipse Modeling Project* (Eclipse Modeling Project, 2010) which provide means to express model-to-model and model-to-code transformations. They import standardized textual *XMI* representations of the models and can parse them according to the used meta-model. Thus, one can easily introduce domain-specific concepts to forward information from a model-level to the model transformations and code generators. Tools like *Papyrus* (PAPYRUS UML, 2011) allow for a graphical representation of the various models and can export them into the *XMI* format. Overall, there is a complete toolchain for graphical modelling and transformation steps available that can be tailored to the domain specific needs of robotics.

MDSD is much more than code generation for different platforms to address the technology change problem and to make development more efficient by automatically generating repetitive code. The benefits of *MDSD* are manifold (Stahl & Völter, 2006; Völter, 2006): (i) models are free of implementation artefacts and directly represent reusable domain knowledge including best practices, (ii) domain experts can play a direct role and are not requested to translate their knowledge into software representations, (iii) design patterns, sophisticated & optimized software structures and approved software solutions can be made available to domain experts and enforced by embedding them in templates for use by highly optimized code generators such that even novices can immediately take advantage from a coded immense experience, (iv) parameters and properties of components required for system level composition and the adaptation to different target systems are explicated and can be modified within a model-based toolchain.

## 2.4 Stable structures and freedom from choice

In robotics, we believe that the cornerstone is a *component model* based on *service-orientation* for its provided and required interactions represented in an abstract way in form of *models*.

A *robotics component model* needs to provide *component level* as well as *system level* concepts, structures and building blocks to support separation of concerns while at the same time ensuring composability based on a composition theory: (i) building blocks out of which one composes a component, (ii) patterns of how to design a well-formed component to achieve system level conformance, (iii) guidance towards providing a suitable granularity of services, (iv) specification of the behavior of interactions and (v) best practices and solutions of domain-specific problems. *MDSD* can then provide toolchains and thereby support separation of concerns and separation of roles.

The above approach asks for the identification of *stable structures* versus *variation points* (Webber & Gomaa, 2004). A robotics component model has to provide guidance via stable structures where these are required to support separation of concerns and to ensure system level conformance. At the same time, it has to allow for freedom wherever possible. The distinction between stable structures and variation points is of relevance at all levels (operating system interfaces, library interfaces, component internal structures, provided and required services etc.). In fact, identified and enforced stable structures come along with restrictions. However, one has to notice that well thought out limitations are not a universal negative and *freedom from choice* (Lee & Seshia, 2011) gives guidance and assurance of properties beyond one's responsibilities in order to ensure separation of concerns.

As detailed in (Schlegel et al., 2011), stable structures with respect to a service-oriented component-based approach can be identified. These are illustrated in figure 6.

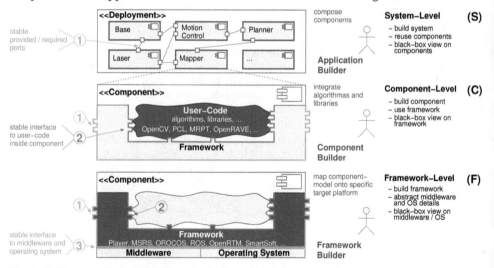

Fig. 6. Stable structures and different roles in a component-based software approach.

At the *system level (S)*, *provided* and *required* service ports ① of a component form a stable interface for the *application builder*. In an ideal situation, all relevant properties of a component are made explicit to support a black box view. Hence, system level properties like resource conformance of the component mapping to the computing platform can be checked during system composition and deployment.

At the *component level (C)*, the *component builder* wants to rely on a stable interface to the component framework ②. In an ideal situation, the component framework can be considered as black box hiding all operating system and middleware aspects from the user code. The component framework adds the execution container to the user code such that the resulting component is conformant to a black box component view.

At the *framework level (F)*, two stable interfaces exist: (i) between the framework and the user code of the component builder ② and (ii) between the framework and the underlying middleware & operating system ③. The stable interface ② ensures that no middleware and operating system specifics are unnecessarily passed on to the component builder. The stable interface ③ ensures that the framework can be mapped onto different implementational technologies (middleware, operating systems) without reimplementing the framework in its entirety. The *framework builder* maintains the framework which links the stable interfaces ② and ③ and maps the framework onto different implementational technologies via the interface ③.

## 3. The SMARTSOFT-approach

The basic idea behind SMARTSOFT (Schlegel, 2011) is to master the component hull and thereby achieve *separation of concerns* as well as *separation of roles*. Figure 7 illustrates the SMARTSOFT component model and how its component hull links the stable interfaces ①, ② and ③.

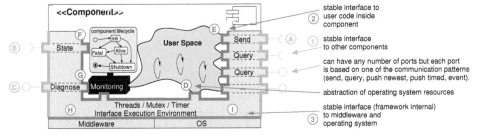

Fig. 7. The structure of a SMARTSOFT component and its stable interfaces.

| Pattern | Description | Service | Description |
|---------|-------------|---------|-------------|
| send | one-way communication | param | component configuration |
| query | two-way request | state | activate/deactivate component services |
| push newest | 1-to-n distribution | wiring | dynamic component wiring |
| push timed | 1-to-n distribution | diagnose | introspection of components |
| event | asynchronous notification | *(internally based on communication patterns)* | |

Table 2. The set of patterns and services of SMARTMARS.

The link between ① and ② is realized by *communication patterns*. Binding a communication pattern with the type of data to be transmitted results in an externally visible service represented as port. The small set of generic and predefined communication patterns listed in the left part of table 2 are the only ones to define externally visible services. Thus, the behavior and usage of a service is immediately evident as soon as one knows its underlying communication pattern.

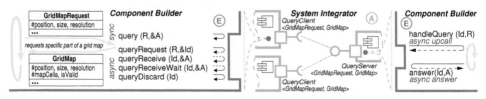

Fig. 8. The views of a component builder and a system integrator on services by the example of a grid map service based on a query communication pattern.

Figure 8 illustrates this concept by means of the *query* communication pattern which consists of a *query client* and a *query server*. The query pattern expects two *communication objects* to define a service: a request object and an answer object. Communication objects are transmitted *by-value* to ensure decoupling of the lifecycles of the client side and the server side of a service. They are arbitrary objects enriched by a unique identifier and get/set-methods. Hidden from the user and inside the communication patterns, the content of a communication object provided via E gets extracted and forwarded to the middleware interface H. Incoming content at H is put into a new instance of the according communication object before providing access to it via E.

In the example, the system integrator sees a provided port based on a *query server* with the communication objects *GridMapRequest* and *GridMap*. The map service might be provided by a map building component. Each component with a port consisting out of a *query client* with the same communication objects can use that service. For example, a path planning component might need a grid map and expose a required port for that service. The *GridMapRequest* object provides the parameters of the individual request (for example, the size of the requested map patch, its origin and resolution) and the *GridMap* returns the answer. The answer is self-contained comprising all the parameters describing the provided map. That allows to interpret the map independently of the current settings of the service providing component and gives the service provider the chance to return a map as close to but different from the requested parameters in case he cannot handle them exactly.

A component builder uses the stable interface E. In case of the client side of a service based on the query pattern, it always consists of the same synchronous as well as asynchronous access modes independent from the used communication objects and the underlying middleware. They can be used from any number of threads in any order. The server side in this example always consists of an asynchronous handler upcall for incoming requests and a separated answer method. This separation is important since it does not require the upcall to wait until the answer is available before returning. We can now implement any kind of processing model inside a component, even a processing pipeline where the last thread calls the answer method, without blocking or wasting system resources of the upcall or be obliged to live with the threading models behind the upcall.

In the example, the upcall at the service provider either directly processes the incoming *GridMapRequest* object or forwards it to a separate processing thread. The requested map patch is put into a *GridMap* object which then is provided as answer via the *answer* method.

It can be seen that the client side is not just a proxy for the server side. Both sides of a communication pattern are completely standalone entities providing stable interfaces A and E by completely hiding all the specifics of H and I (see figure 7). One can neither expose arbitrary member functions at the outside component hull nor can one dilute the semantics and behavior of ports. The different communication patterns and their internals are explained in detail in (Schlegel, 2007).

Besides the services defined by the component builder *(A)*, several predefined services exist to support system level concerns (Lotz et al., 2011). Each component needs to provide a *state service* to support system level orchestration (outside view *B*: activation, deactivation, reconfiguration; inside view *F*: manage transitions between service activations, support housekeeping activities by entry/exit actions). An optional *diagnostic service (C, G)* supports runtime monitoring of the component. The optional *param service* manages parameters by name/value-pairs and allows to change them at runtime. The optional *wiring service* allows to wire required services of a component at runtime from outside the component. This is needed for task and context dependent composition of behaviors.

### 3.1 The SMARTMARS meta-model

All the stable interfaces, concepts and structures as well as knowledge about which ingredients and structures form a well-formed SMARTSOFT component and a well-formed system of SMARTSOFT components are explicated in the SMARTMARS meta-model (figure 9). The meta-model is abstract, universally valid and independent from implementation technologies (e.g. UML profile (Fuentes-Fernández & Vallecillo-Moreno, 2004), eCore (Gronback, 2009)). It provides the input for tool support for the different roles (like component developer, system integrator etc.), explicates separation of concerns and can be mapped onto different software technologies (e.g. different types of middleware like CORBA, ACE (Schmidt, 2011) and different types of operating systems).

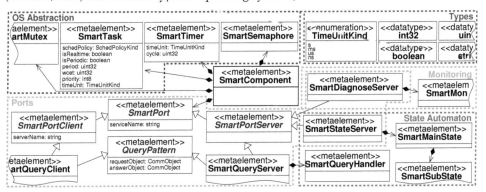

Fig. 9. Excerpt of the SMARTMARS meta-model.

### 3.2 Policies and strategies behind SMARTSOFT services

A major part of the SMARTSOFT approach are the policies and strategies that manifest themselves in the structure of the component model, explain its building blocks and guide their usage.

Separation of the roles of a component developer and a system integrator requires to control the interface between the inner part of a component and its outer part and to control this boundary. As soon as one gains control over the component hull, one can make sure that all relevant properties and parameters needed for the black box view of the system integrator become explicated at the component hull. One can also make sure that a component developer has no chance to expose component internals to the outside. SMARTSOFT achieves this via predefined communication patterns as the only building blocks to define externally visible services and by further guidelines on how to build good services.

*A basic principle is that clients of services are not allowed to make any assumptions about offered services beyond the announced characteristics and that service providers are not allowed to make any assumptions about service requestors (like e.g. their maximum rate of requests).*

This principle results in simple and precise guidelines of how to apply the communication patterns in order to come up with well-formed services. As long as a service is being offered, the service provider has to accept all incoming requests and has to respond to them according to its announced quality-of-service parameters.

We illustrate this principle by means of the *query pattern*. As long as there are no further quality-of-service attributes, the service provider accepts all incoming requests and guarantees to answer all accepted requests. However, only the service provider knows about its resources available to process incoming requests and clients are not allowed to impose constraints on the service provider (a request might provide further non-committal hints to the service provider like a request priority). Thus, the service provider is allowed to provide a nil answer (the flag *is valid* is set to false in the answer) in case he is running out of resources to answer a particular request. In consequence, all service requestors always must be prepared to get a nil answer. A service requestor is also not allowed to make any assumptions about the response time as long as according quality-of-service attributes are not set by the service provider. However, if a service provider announces to answer requests within a certain time limit, one can rely on getting at least a nil answer before the deadline. If a service requestor depends on a maximum response time although this quality-of-service attribute is not offered by the service provider, he needs to use client-side timeouts with his request. This overall principle ensures (i) loose coupling of services, (ii) prevents clients from imposing constraints on service providers and (iii) gives service providers the means to arbitrate requests in case of limited resources.

It now also becomes evident why SMARTSOFT offers more than just a request/response and a publish/subscribe pattern which would be sufficient to cover all communicational needs. The *send* pattern explicates a one-way communication although one can emulate it via a query pattern with a void answer object. However, practical experience proved that a much better clarity for services with this characteristic is achieved when offering a separate pattern. The same holds true for the *push newest* and the *push timed* pattern. In principle, the push timed pattern is a push newest pattern with a regular update. However, in case of a push newest pattern, service requestors rely on having the latest data available at any time. This is different from a push timed pattern where the focus is on the service provider guaranteeing a regular time interval (in some cases even providing the same data). Although one could cover some of these aspects by quality-of-service attributes, they also have an impact on the kind of perception of its usage by a component developer. Again, achieving clarity and making the characteristics easily recognizable is of particular importance for the strict separation of the roles of component developers and system integrators. This also becomes obvious with the *event* pattern. In contrast to the push patterns, service requestors get informed only in case a server side event predicate (service requestors individually parametrize each event activation) becomes true. This tremendously saves bandwidth compared to publishing latest changes to all clients since one then always would have to publish a snapshot of the overall context needed to evaluate the predicate at the client side instead of just the information when an event fired.

### 3.3 A robotics example illustrating the SMARTSOFT concepts

Figure 10 illustrates how the SMARTSOFT component model and its meta-elements provided by SMARTMARS structure and partition a typical robotics use-case, namely the navigation of a mobile platform. Besides access to sensor data and to the mobile base, algorithmic building blocks of a navigation system are map building, path planning, motion execution and self localization. Since these building blocks are generic for navigation systems independently of the used algorithms, it makes sense to come up with an according component structure and services (or expect readily available components and services).

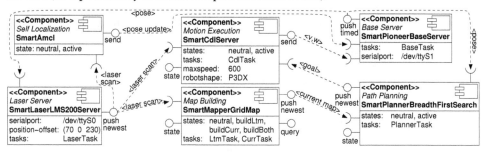

Fig. 10. Structure of a navigation task based on the SMARTSOFT component model.

The SmartLaserLMS200Server component provides the latest laser scan via a push newest port. Thus, all subscribed clients always get an update as soon as a new laserscan is available. It is subscribed to the pose service of the robot base to label laser scans with pose stamps. The component comprises a SmartTask to handle the internal communication with the laser hardware. This way, the aliveness of the overall component and its services is not affected by flaws on the laser hardware interface. Parameters like position-offset and serialport are used to customize the component to the target robotic system. These parameters have to be set by the application builder during the deployment step. The SmartMapperGridMap component requires a laser scan to build the longterm and the current map. The current map is provided by a push newest server port (as soon as a new map is available, it is provided to subscribed clients which makes sense since path planning depends on latest maps) and the longterm map by a query server port (since it is not needed regularly, it makes sense to provide it only on a per-request basis). The state port is used to set the component into different states depending on which services are needed in the current situation: build no map at all (neutral), build the current map only (buildCurr), build the longterm map only (buildLtm) or build both maps (buildBoth). The push newest server publishes the current map only in the states buildCurr and buildBoth. Requests for a longterm map are answered as long as the component and its services are alive but with an invalid map in case it is in the states neutral or buildCurr (valid flag of answer object set to false). Accordingly, the SmartPlannerBreadthFirstSearch component provides its intermediate waypoints by a push newest server (update the motion execution component as soon as new information is available). The motion execution component regularly commands new velocities to the robot base via a send service. The motion execution component is also subscribed to the laser scan service to be able to immediately react to obstacles in dynamic environments. This way, the different services interact to build various control loops to combine goal directed and reactive navigation while at the same time allowing for replacement of components.

### 3.4 State-of-the-art and related work

The historical need in robotics to be responsible for creation of the application logic and to be at the same time the system integrator generated a poor understanding in the robotics community that these two roles ought to be separated.  In consequence, most robotics frameworks don't make this distinction and consequently they don't offer any clear guideline to the developer on how to achieve separation of roles.

For example, *ROS* (Quigley et al., 2009) is a currently widely-used framework in robotics providing a huge and valuable codebase.  However, it lacks guidance for component developers to ensure system level conformance for composability.  Instead, its focus is on side-by-side existence of all kinds of overlapping concepts without an abstract representation of its core features and properties in a way independent of any implementation.

The only approach in line with the presented concepts is the *RTC Specification* (OMG, 2008) which is considered the most advanced concept of *MDSD* in robotics. However, it is strongly influenced by use-cases requiring a data-flow architecture and they do not yet considerably take into account requirements imposed by runtime adaptability.

## 4. Reference implementation of the SMARTMDSD TOOLCHAIN

The reference implementation of the SMARTMDSD TOOLCHAIN implements the SMARTMARS meta-model within a particular MDSD-toolchain.  It is used in real world operation to develop components and to compose complex systems out of them. The focus of this section is on technical details of the implementation of a meta-model. Another focus is on the role-specific view and the support a MDSD-toolchain provides. We illustrate the reference implementation of the toolchain along the different roles of the stakeholders and their views on the toolchain.

### 4.1 Decisions and tools behind the reference implementation - framework builder view

The reference implementation of our SMARTMDSD TOOLCHAIN is based on the *Eclipse Modeling Project (EMP)* (Eclipse Modeling Project, 2010) and *Papyrus UML* (PAPYRUS UML, 2011).

*Papyrus UML* is used as graphical modeling tool in our toolchain. Therefore, it is customized by the framework builder for the development of SMARTSOFT Components (component builder) and deployments of components (application builder). This includes for example a customized wizard to create communication objects, components as well as deployments. The modeling view of *Papyrus UML* is enriched with a customized set of meta-elements to create the models. The model transformation and code generation steps are developed with *Xpand* and *Xtend* (Efftinge et al., 2008) which are part of the *EMP*. These internals are not visible to the component builder and the application builder. They just see the graphical modeling tool to create their models and use the *CDT Eclipse Plugin* (Eclipse CDT, 2011) to extend the source code and to compile binaries. The SMARTMARS meta-model is implemented as a *UML Profile* (Fuentes-Fernández & Vallecillo-Moreno, 2004) using *Papyrus UML*.

The decision to use *UML Profiles* and *Papyrus UML* to implement our toolchain is motivated by the reduced effort to come up with a graphical modeling environment customized to the robotics domain and its requirements by reusing available tools from other communities. Although some shortcomings have to be accepted and taken into account we were not caught in the huge effort related to implementing a full-fledged *GMF*-based development environment. This allowed us to early come up with our toolchain and to gain deeper insights and more experience on the different levels of abstraction. However, the major drawbacks of *UML Profiles* are:

- *UML* is a general purpose modeling language covering aspects of several domains and is thus complex. Using profiles, it is only possible to enrich *UML*, but not to remove elements.

- Deployment and instantiations of components are not adequately supported.

- *UML Profiles* provide just a lightweight extension of *UML*. That means, the structure of *UML* itself cannot be modified. The elements can be customized only by stereotypes and tagged values.

To counter the drawbacks of *UML Profiles*, we only support the usage of the stereotyped elements provided by SMARTMARS to create the models of the components and deployments. Directly using pure *UML* elements in the diagrams is not supported. Thus, the models are created using just the meta-elements provided by SMARTMARS. Restricting the usage to SMARTMARS meta-elements, a mapping to another meta-model implementation technology like *eCore* (Gronback, 2009) is straightforward. The stereotyped elements can be mapped onto *eCore* without taking into account *UML* and its structure. In the current implementation of our toolchain, the restriction to only use SMARTMARS meta-elements is enforced with *check* (Efftinge et al., 2008), the *EMP* implementation of *OCL* (Object Management Group, 2010). In the model transformation and code generation steps of our toolchain pure *UML* elements are ignored. Another approach would be to customize the diagrams by removing the *UML* elements from the palette (see fig. 12) and thus restricting their usage. The latter approach is on the agenda of the *Papyrus UML* project and will be supported by future releases.

### 4.2 Development of components – component builder view

Figure 11 illustrates the roles of the framework builder and the component builder. The component builder creates a model of the component using the Eclipse based toolchain, focusing on the component hull. Pushing the button he receives the source files where to integrate the business logic (algorithms, libraries) of the component. During this process the component builder is supported and guided by the toolchain. The internals of the model transformation and code generation steps implemented by the framework builder are not visible to the component builder.

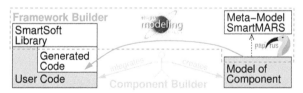

Fig. 11. The component builder models a component, gets the source code of its overall structure (component hull, tasks, etc.) generated by the toolchain and can then integrate user-code into these structures.

The view of the component builder on the toolchain is depicted in figure 12. It is illustrated by a face recognition component which is a building block in many service robotics scenarios as part of the human-robot interface (detection, identification and memorization of persons). In its active state, the component shall receive camera images, apply face recognition algorithms and report detected and recognized persons. Thus, besides the standard ports for setting states (active, neutral) and parameters, we need to specify a port to receive the latest camera images (based on a push newest client) and another one to report on the results (based on an event server). The component shall run the face recognition based on a commercially available library within one thread and optional visualization mechanisms within a second and separated thread. Thus, we need to specify two tasks within the component.

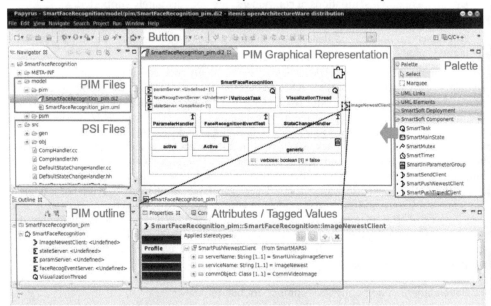

Fig. 12. Screenshot of our toolchain showing the view of the Component Builder.

To create the model the component builder uses the SMARTMARS meta-elements offered in the *palette*. The elements of the created model can be accessed either in the outline view or directly in the graphical representation. Several of the meta-element attributes (tagged values) can be customized and modified in the properties tab (e.g. customizing services to ports, specifying properties of tasks, etc.). The model is stored in files specific to *Papyrus UML*. Pushing the button, the workflow is started and the *PSI* (source) files are generated. The user code files are directly accessible in the *src* folder. The component builder integrates his business logic into these files (in our example, the interaction with the face recognition library). The generated files the component builder must not modify are stored in the *gen* folder. These files are generated and overwritten each time the workflow is executed. For the further processing of the source files, the *Eclipse CDT plugin* is used (*Makefile Project*). The makefile is also generated by the workflow specific to the model properties. User modifications in the makefile can be done inside of *protected regions* (Gronback, 2009).

### 4.3 Development of components – framework builder view

Taking a look behind the scenes of the toolchain, the workflow (fig. 13) appears as a two step transformation according to the *OMG MDA* (Object Management Group & Soley, 2000). The *Platform Independent Model (PIM)*, which is created by the component builder using the meta-elements provided by the *PIM UML Profile*, specifies the component independently of the implementation technology. The first step in the workflow is the model-to-model

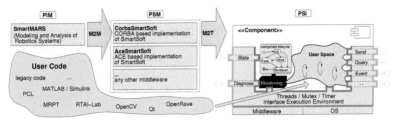

Fig. 13. Two step transformation workflow: Framework Builder view.

(M2M) transformation (encoded with *Xtend*) from the *PIM* into a *Platform Specific Model (PSM)*. In this step the elements of the *PIM* are transformed into corresponding elements of the *PSM* according to the selected target platform. The second step is the model-to-text (M2T) transformation (encoded with *Xpand* and *Xtend*) from the *PSM* into a *Platform Specific Implementation (PSI)*. This transformation is based on customizable code templates.

### 4.3.1 The SmartMARS UML profiles (PIM/PSM)

The abstract SMARTMARS meta-model is implemented by the framework builder as *UML Profile* using *Papyrus UML*. Therefore, standard *UML* elements (e.g. *Component*, *Class*, *Port*) are extended by stereotypes (e.g. SMARTCOMPONENT, SMARTTASK, SMARTQUERYSERVER) to give the meta-elements a new meaning according to the SMARTMARS concept. To distinguish and highlight the new element, it has its own icon attached. Tagged values are used to enrich the meta-element by new attributes which are not provided by the base *UML* element.

In fact there are two *UML Profiles*: one for the *PIM* and one for the *PSM*. The *PIM UML Profile* is visible to the component builder and is used by him to create the models of the components. For each SMARTSOFT implementation (e.g. CORBA, ACE), a *PSM UML Profile* has to be provided covering the specifics of the implementation. For example, the CORBA-based *PSM* supports *RTAI* linux to provide hard realtime tasks. This is represented by the meta-element *RTAITask*. The *PSM UML Profile* is not visible to the component builder and only used by the transformation steps inside the toolchain.

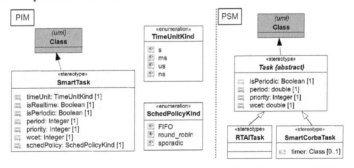

Fig. 14. Screenshots of excerpt of the UML Profiles created with *Papyrus UML* showing the metaelements dedicated to the SMARTTASK. Left: *PIM*; Right: *PSM* with the two variants (1) standard task and (2) *RTAI* task.

An excerpt of the *UML Profiles* is illustrated in figure 14. In the *UML Profile* for the *PIM*, the SMARTTASK extends the *UML class* and enriches it with attributes (tagged values) like *isPeriodic*, *isRealtime*, *period* and *timeUnit*. For the *timeUnit* an enumeration (*TimeUnitKind*) is used to specify the unit in which time values are annotated. In the *UML Profile* for the *CORBA-based PSM*, an abstract task is specified (cannot be instantiated) and the two variants (1) standard task and (2) realtime task are derived from it. They are both not abstract and can thus be instantiated by the component builder to create the model. The standard task adds an optional attribute referencing to a SMARTTIMER meta-element. This is used to emulate periodic non-realtime tasks which are not natively supported by standard tasks of the *CORBA*-based SMARTSOFT implementation.

### 4.3.2 Model transformation and code generation steps

The *M2M* transformation maps the platform independent elements of the *PIM* onto platform specific elements of the selected target platform. Such a mapping is illustrated by the example of the *SmartTask* (fig. 15 *left*) and the *CORBA*-based *PSM*. The SMARTTASK comprises several elements which are necessary to describe a task behavior and its characteristics.

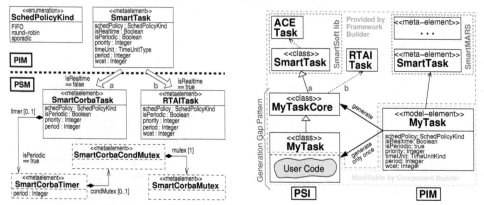

Fig. 15. Model transformation and code generation steps illustrated by the example of the SMARTTASK. *Left:* Transformation of the PIM into a PSM. *Right:* Code generation and Generation Gap Pattern.

```
task_mutex.ext
create uml::Class this addSmartTask(SmartMARS::SmartTask tsk, uml::Component cmp) :
    cmp.packagedElement.add(this) ->
    this.setName(tsk.name) ->
    if( tsk.isRealtime == true) then
    {
        this.applyStereotype("CorbaSmartSoft::RTAITask") ->
        setTaggedValue(this, "CorbaSmartSoft::RTAITask", "isPeriodic", tsk.isPeriodic) ->
        setTaggedValue(this, "CorbaSmartSoft::RTAITask", "wcet", tsk.wcet.toSecond(tsk.timeUnit.name)) ->
        setTaggedValue(this, "CorbaSmartSoft::RTAITask", "period", tsk.period.toSecond(tsk.timeUnit.name)) ->
        setTaggedValue(this, "CorbaSmartSoft::RTAITask", "priority", tsk.priority)
    }
    else
    {
        this.applyStereotype("CorbaSmartSoft::SmartCorbaTask") ->
        setTaggedValue(this, "CorbaSmartSoft::SmartCorbaTask", "isPeriodic", tsk.isPeriodic) ->
        setTaggedValue(this, "CorbaSmartSoft::SmartCorbaTask", "wcet", tsk.wcet.toSecond(tsk.timeUnit.name)) ->
        setTaggedValue(this, "CorbaSmartSoft::SmartCorbaTask", "period", tsk.period.toSecond(tsk.timeUnit.name)) ->
        setTaggedValue(this, "CorbaSmartSoft::SmartCorbaTask", "priority", tsk.priority) ->
        if( tsk.isPeriodic == true ) then
        {
            setTaggedValue(this, "CorbaSmartSoft::SmartCorbaTask", "timer", cmp.addTimer(tsk.name, tsk.period, tsk.timeUnit.name))
        }
    };
```

Fig. 16. PIM to PSM model transformation of the SMARTTASK depending on the attribute *isRealtime*.

```
smartTask.xpt     PSM → PSI Template

«DEFINE TaskUserSourceFile FOR CorbaSmartSoft::Task-»
«FILE this.getUserSourceFilename() writeOnce-»
«getCopyrightWriteOnce()»
#include "«this.getUserHeaderFilename()»"
#include "gen/«((CorbaSmartSoft::SmartCorbaComponent)this.eContainer()).getCoreHeaderFilename()»"

#include <iostream>

«this.getName()»::«this.getName()»()
{
    std::cout << "constructor «this.getName()»\n";
}

int «this.getName()»::svc()
{
    // do something -- put your code here !!!
    while(1)
    {
        «IF this.isPeriodic == true-»
        std::cout << "Hello from «this.getName()»  - periodic\n";
        smart_task_wait_period();
        «ELSE-»
        std::cout << "Hello from «this.getName()»\n";
        sleep(1);
        «ENDIF-»
    }
    return 0;
}
«ENDFILE»
«ENDDEFINE»
```

```
ServoTask.cc     PSI (user code .cc file)

#include "ServoTask.hh"
#include "gen/SmartServo.hh"

#include <iostream>

ServoTask::ServoTask()
{
    std::cout << "constructor ServoTask\n";
}

int ServoTask::svc()
{
    // do something -- put your code here !!!
    while(1)
    {
        std::cout << "Hello from ServoTask  - periodic\n";
        smart_task_wait_period();
    }
    return 0;
}
```

Fig. 17. *PSM* to *PSI* transformation of the SMARTTASK. *Left:* Excerpt of the transformation template (*xPand*) generating the PSI of a standard task. *Right:* The generated code where the user adds the business logic of the task.

Depending on the attribute *isRealtime* the SMARTTASK is either mapped onto a RTAITASK or a non-realtime SMARTCORBATASK[1]. The *Xtend* transformation rule to transform the *PIM* SMARTTASK into the appropriate *PSM* element is depicted in figure 16.

In case the attributes specify a non-realtime, periodic SMARTTASK, the toolchain extends the *PSM* by the elements needed to emulate periodic tasks (as this feature is not covered by standard tasks). In each case the user integrates his algorithms and libraries into the stable interface provided by the SMARTTASK (component builder view) independent of the hidden internal mapping of the SMARTTASK (generated code). Figure 17 depicts the *Xpand* template to generate the user code file for the task in the *PSI*. The figure shows the template on the left and the generated code on the right.

The *PSI* consists of the SMARTSOFT library, the generated code and the user code (fig. 15 *right*). To be able to re-generate parts of the component source code according to modified parameters in the model without affecting the source code parts added by the component builder, the generation gap pattern (Vlissides, 2009) is used. It is based on inheritance – the user code inherits from the generated code[2]. The source files called *generated code* are generated each time the transformation workflow in the toolchain is executed. These files contain the logic which is generated behind the scenes according to the model parameters and must not be modified by the component builder. The source files called *user code* are just generated if they do not already exist. They are intended for the component builder to add the algorithms and libraries. The generation of the user code files is more for the convenience of the component builder to have a code template as starting point. These files are in the full responsibility of the component builder and are never modified or overwritten by the transformation workflow of the toolchain. In this context *generate once* means that the file is only generated if it does not already exist. This is typically the case if the workflow is executed for the first time. The clear separation of generated code and user code by the generation gap pattern allows on the one hand to reflect modifications of the model in the generated source

---

[1] *Corba* in element names indicates that the element belongs to the *CORBA* specific *PSM*.

[2] The pattern could also be used in the opposite inheritance ordering so that the generated code inherits from the user code.

code without overwriting the user parts. On the other hand it gives the user the freedom to structure his source code according to his needs and does not restrict the structure as would be the case with, for example, *protected regions*. Consequently, the component builder can modify the *period*, *priority* or even the *isRealtime* attribute of the task in the model, re-generate and compile the code without requiring any modification in the user code files. The modification in the model just affects the generated code part of the *PSI*.

## 4.4 Deployment of components – application builder view

The deployment is used to compose a robotic system out of available components. The application builder imports the desired components and places them onto the target platform. Furthermore, he defines the initial wiring of the components by connecting the ports with the meta-element *Connection*. Figure 18 illustrates the composition of navigation

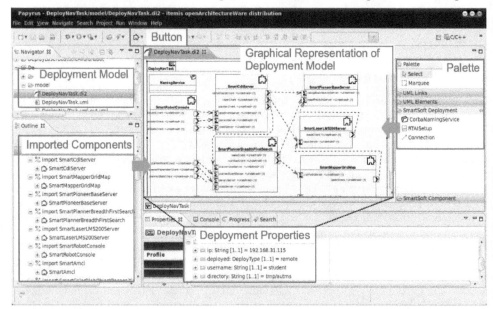

Fig. 18. Screenshot of our toolchain showing the deployment of components to build a robotic system.

components. In this example, the application builder (system integrator) imports components specific to a particular robot platform (SmartPioneerBaseServer) and specific to a particular sensor (SmartLaserLMS200Server). The navigation components (SmartMapperGridMap, SmartPlannerBreadthFirstSearch, SmartCDLServer) can be used across different mobile robots. The SmartRobotConsole provides a user interface to command the robot.

The components are presented to the application builder as black boxes with dedicated variation points. These have to be bound during the deployment step and can be specified according to system level requirements. For example, a laser ranger component might need the coordinates of its mounting point relative to the robot coordinate system. One might also reduce the maximum scanning frequency to save computing resources. Parameters also need to be bound for the target system. For example, in case *RTAI* is used inside of a component, the *RTAI* scheduler parameters (timer model underlying RTAI: periodic, oneshot) of the target

*RTAI* system have to be specified. If the application builder forgets to bind required settings, this absence is reported to him by the toolchain.

The application builder can identify the provided and required services of a component via its ports. He can inspect its characteristics by clicking on the port icon which opens a property view. That comprises the communication pattern type, the used communication objects and further characteristics like service name and also port specific information like update frequencies. The initial wiring is done within the graphical representation of the model. In case the application builder wants to connect incompatible ports, the toolchain refuses the connection and gives further hints on the reasons.

If the *CORBA*-based implementation of SMARTSOFT is used, the *CORBA* naming service properties *IP*-address and *port*-number have to be set. Furthermore, the deployment type (*local, remote*) has to be selected. For a remote deployment, the *IP*-address, *username* and *target folder* of the target computer have to be specified. The deployed system is copied to the target computer and can be executed there. In case of a local deployment, the system is customized to run on the local machine of the application builder. This is, for example, the case if no real robot is used and the deployed system uses simulation components (e.g. *Gazebo*). Depending on the initial wiring, parameter files are generated and also copied into the deployment folder. These parameter files contain application specific adjustments of the components. In addition, a shell script to start the system is generated out of the deployment model.

### 4.5 Deployment of components – framework builder view

To implement the deployment of components, some meta-elements are added by the framework builder to the *UML Profile* (fig. 19). This section focuses on the *CORBA*-based deployment.

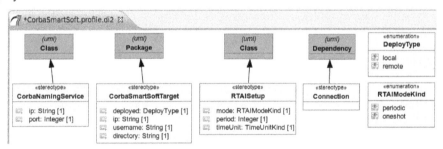

Fig. 19. Meta-elements to support the deployment of components.

The deployment model contains relevant information like the initial wiring between components (*Connection*), naming service properties (*CorbaNamingService*), scheduler properties (*RTAISetup*) and parameters about the deployment itself (*CorbaSmartSoftTarget*). The models of the components are made available to the deployment model using the *UML import* mechanism. This allows to access the internal structure of the components. Out of the deployment model the parameter files and a start script are generated (*M2T*) using *Xpand* and *Xtend* in a similar way as these transformation languages are used to generate code for the components. Based on the deployment model several analysis and simulation models can be generated to get feedback from 3rd-party tools. For example, one can extract parameters of all realtime tasks mapped onto a specific processor to perform hard realtime schedulability analysis (CHEDDAR (Cheddar, 2010)) (Schlegel et al., 2010).

Fig. 20. The clean-up scenario. (1) Kate approaches the table; (2/3) Kate stacks objects into each other; (4) Kate throws cups into the kitchen sink.

As deployments and especially instantiations of components are not sufficiently supported by *UML*, a few workarounds are necessary as long as the SMARTMARS meta-model is implemented as *UML Profile*. For example, a robot with two laser range finders (front, rear) requires two instances of the same component. Each laser instance requires its individual parameters (e.g. serial port, pose on robot). These parameters are assigned to the deployment model by the application builder specifically for each component. In the implementation based on the *UML Profile*, we hence work on copies of components. Individual instances with their own parameter sets are considered in the abstract SMARTMARS meta-model and are also covered in the SMARTSOFT implementation, thus switching to a different meta-model implementation technology would allow for instances. This has not yet been done due to the huge manpower needed compared to just reusing *UML* tools.

## 5. Example / scenario

The work presented has been used to build and run several real-world scenarios, including the participation at the RoboCup@Home challenge. Among other tasks our robot "Kate" can follow persons, deliver drinks, recognize persons and objects and interact with humans by gestures and speech.

In the clean-up scenario [3] (fig. 20) the robot approaches a table, recognizes the objects which are placed on the table and cleans the table either by throwing the objects into the trash bin or into the kitchen sink. There are different objects, like cups, beverage cans and different types of crisp cans. The cups can be stacked into each other and have to be thrown into the kitchen sink. Beverage cans can be stacked into crisp cans and have to be thrown into the trash bin. Depending on the type of crisp can, one or two beverage cans can be stacked into one crisp can. After throwing some of the objects into the correct disposal the robot has to decide whether to drive back to the table to clean up the remaining objects (if existing) or to drive to the operator and announce the result of the cleaning task. The robot reports whether all objects on the table could be cleaned up or, in case any problems occurred, reports how many objects are still left.

Such complex and different scenarios can neither be developed from scratch nor can their overall system complexity be handled without using appropriate software engineering methods. Due to their overall complexity and richness, they are considered as convincing stress test for the proposed approach. In the following the development of the cleanup example scenario is illustrated according to the different roles.

The **framework builder** provides the tools to develop SMARTSOFT components as well as to perform deployments of components to build a robotic system. In the described example this includes the *CORBA*-based implementation of the SMARTSOFT framework

---

[3] http://www.youtube.com/roboticsathsulm#p/u/0/xtLK-655v7k

and the SMARTMDSD toolchain which are both available on *Sourceforge* (http://smart-robotics.sourceforge.net).

The component builder view of the SMARTMDSD toolchain supports **component builders** to develop their components independently of each other, but based on agreed interfaces. These components are independent of the concrete implementation technology of SMARTSOFT. Component builders provide their components in a component shelf. The models of the components include all information to allow a black-box view of the components (e.g. services, properties, resources). The explication of such information about the components is required by the application builder to compose robotic systems in a systematic way. To orchestrate the components at run-time, the task coordination language SMARTTCL (Steck & Schlegel, 2010) is used. Therefore, SMARTTCL is wrapped by a SMARTSOFT component and is also provided in the component shelf. The SMARTTCL component provides reusable action plots which can be composed and extended to form the desired behavior of the robot.

The **application builder** uses the application builder view of the SMARTMDSD toolchain. He composes already existing components to build the complete robotic system. In the above described cleanup scenario, 17 components (e.g. mapping, path planning, collision avoidance, laser ranger, robot-base) are reused from the component shelf. It is worth noting that the components were not particularly developed for the cleanup scenario, but can be used in the cleanup scenario due to the generic services they provide. The SMARTTCL sequencer component is customized according to the desired behavior of the cleanup scenario. Therefore, several of the already existing action plots can be reused. Application specific extensions are added by the application builder.

At run-time the SMARTTCL sequencer component coordinates the software components of the **robot** by modifying the configuration and parametrization as well as the wiring between the components. As SMARTTCL can access the information (e.g. parameters, resources) explicated in the models of the components at run-time, this information can be taken into account by the decision making process. That allows the robot not only to take the current situation and context into account, but also the configuration and resource usage of the components. In the described scenario, the sequencer manages the resources of the overall system, for example, by switching off components which are not required in the current situation. While the robot is manipulating objects on the table and requires all available computational resources for the trajectory planning of the manipulator, the components for navigation are switched off.

## 6. Conclusion

The service-oriented component-based software approach allows separation of roles and is an important step towards the overall vision of a robotics software component shelf. The feasibility of the overall approach has been demonstrated by an Eclipse-based toolchain and its application within complex Robocup@Home scenarios. Next steps towards model-centric robotic systems that comprehensively bridge design-time and runtime model usage now become viable.

## 7. References

Andrade, L., Fiadeiro, J. L., Gouveia, J. & Koutsoukos, G. (2002). Separating computation, coordination and configuration, *Journal of Software Maintenance* 14(5): 353–369.
Beydeda, S., Book, M. & Gruhn, V. (eds) (2005). *Model-Driven Software Development*, Springer.

Björkelund, A., Edström, L., Haage, M., Malec, J., Nilsson, K., Nugues, P., Robertz, S. G., Störkle, D., Blomdell, A., Johansson, R., Linderoth, M., Nilsson, A., Robertsson, A., Stolt, A. & Bruyninckx, H. (2011). On the integration of skilled robot motions for productivity in manufacturing, *Proc. IEEE Int. Symposium on Assembly and Manufacturing*, Tampere, Finland.

Blogspot (2008). Discussion of Aspect oriented programming(AOP).
  URL: *http://programmingaspects.blogspot.com/*

Bruyninckx, H. (2011). Separation of Concerns: The 5Cs - Levels of Complexity, Lecture Notes, Embedded Control Systems.
  URL: *http://people.mech.kuleuven.be/ bruyninc/ecs/LevelsOfComplexity-5C-20110223.pdf*

CBDI Forum (2011). CBDI Service Oriented Architecture Practice Portal - Independent Guidance for Service Architecture and Engineering.
  URL: *http://everware-cbdi.com/cbdi-forum*

Cheddar (2010). A free real time scheduling analyzer.
  URL: *http://beru.univ-brest.fr/ singhoff/cheddar/*

Chris, R. (1989). *Elements of functional programming*, Addison-Wesley Longman Publishing Co, Boston, MA.

Delamer, I. & Lastra, J. (2007). Loosely-coupled automation systems using device-level SOA, *5th IEEE International Conference on Industrial Informatics*, Vol. 2, pp. 743–748.

Dijkstra, E. (1976). *A Discipline of Programming*, Prentice Hall, Englewood Cliffs, NJ.

Eclipse CDT (2011). C/C++ Development Tooling for Eclipse.
  URL: *http://www.eclipse.org/cdt/*

Eclipse Modeling Project (2010). Modeling framework and code generation facility.
  URL: *http://www.eclipse.org/modeling/*

Efftinge, S., Friese, P., Haase, A., Hübner, D., Kadura, C., Kolb, B., Köhnlein, J., Moroff, D., Thoms, K., Völter, M., Schönbach, P., Eysholdt, M. & Reinisch, S. (2008). openArchitectureWare User Guide 4.3.1.

Fuentes-Fernández, L. & Vallecillo-Moreno, A. (2004). An Introduction to UML Profiles, *UPGRADE* Volume V(2): 6–13.

Gelernter, D. & Carriero, N. (1992). Coordination languages and their significance, *Commun. ACM* 35(2): 97–107.

Gronback, R. C. (2009). *Eclipse Modeling Project: A Domain-Specific Language (DSL) Toolkit*, Addison-Wesley, Upper Saddle River, NJ.

Heineman, G. T. & Councill, W. T. (eds) (2001). *Component-Based Software Engineering: Putting the Pieces Together*, Addison-Wesley Professional.

IBM (2006). Model-Driven Software Development, *Systems Journal* 45(3).

Lastra, J. L. M. & Delamer, I. M. (2006). Semantic web services in factory automation: Fundamental insights and research roadmap, *IEEE Trans. Ind. Informatics* 2: 1–11.

Lau, K.-K. & Wang, Z. (2007). Software component models, *IEEE Transactions on Software Engineering* 33: 709–724.

Lee, E. A. & Seshia, S. A. (2011). *Introduction to Embedded Systems - A Cyber-Physical Systems Approach*, ISBN 978-0-557-70857-4.
  URL: *http://LeeSeshia.org*

Lotz, A., Steck, A. & Schlegel, C. (2011). Runtime monitoring of robotics software components: Increasing robustness of service robotic systems, *Proc. 15th Int. Conference on Advanced Robotics (ICAR)*, Tallinn, Estland.

Meyer, B. (2000). What to compose, *Software Development* 8(3): 59, 71, 74–75.

Mili, H., Elkharraz, A. & Mcheick, H. (2004). Understanding separation of concerns, *Proc. Workshop Early Aspects: Aspect-Oriented Requirements Engineering and Architecture Design*, Vancouver, Canada, pp. 75–84.

Neelamkavil, F. (1987). *Computer simulation and modeling*, John Wiley & Sons Inc.

Object Management Group (2010). Object Constraint Language (OCL).
    URL: *http://www.omg.org/spec/OCL/*

Object Management Group & Soley, R. (2000). Model-Driven Architecture (MDA).
    URL: *http://www.omg.org/mda*

OMG (2008). Robotic Technology Component (RTC).
    URL: *http://www.omg.org/spec/RTC/*

PAPYRUS UML (2011). Graphical editing tool for uml.
    URL: *http://www.eclipse.org/modeling/mdt/papyrus/*

Parnas, D. (1972). On the criteria to be used in decomposing systems into modules, *Communications of the ACM* 15(12).

Quigley, M., Gerkey, B., Conley, K., Faust, J., Foote, T., Leibs, J., Berger, E., Wheeler, R. & Ng, A. (2009). ROS: An open-source Robot Operating System, *ICRA Workshop on Open Source Software*.

Radestock, M. & Eisenbach, S. (1996). Coordination in evolving systems, *Trends in Distributed Systems – CORBA and Beyond*, Springer-Verlag, pp. 162–176.

Reiser, U., Connette, C., Fischer, J., Kubacki, J., Bubeck, A., Weisshardt, F., Jacobs, T., Parlitz, C., Hägele, M. & Verl, A. (2009). Care-O-bot 3 – Creating a product vision for service robot applications by integrating design and technology, *Proc. IEEE/RSJ International Conference on Intelligent Robots and Systems (ICRA)*, St. Louis, USA, pp. 1992–1997.

Schlegel, C. (2007). Communication patterns as key towards component interoperability, *in* D. Brugali (ed.), *Software Engineering for Experimental Robotics, STAR 30*, Springer, pp. 183–210.

Schlegel, C. (2011). SMARTSOFT – Components and Toolchain for Robotics.
    URL: *http://smart-robotics.sf.net/*

Schlegel, C., Steck, A., Brugali, D. & Knoll, A. (2010). Design abstraction and processes in robotics: From code-driven to model-driven engineering, *2nd Int. Conf. on Simulation, Modeling, and Programming for Autonomous Robots (SIMPAR)*, Springer LNAI 6472, pp. 324–335.

Schlegel, C., Steck, A. & Lotz, A. (2011). Model-driven software development in robotics: Communication patterns as key for a robotics component model, *Introduction to Modern Robotics*, iConcept Press.

Schmidt, D. (2011). The ADAPTIVE Communication Environment.
    URL: *http://www.cs.wustl.edu/ schmidt/ACE.html*

Sprott, D. & Wilkes, L. (2004). CBDI Forum.
    URL: *http://msdn.microsoft.com/en-us/library/aa480021.aspx*

Stahl, T. & Völter, M. (2006). *Model-Driven Software Development: Technology, Engineering, Management*, Wiley.

Steck, A. & Schlegel, C. (2010). SmartTCL: An Execution Language for Conditional Reactive Task Execution in a Three Layer Architecture for Service Robots, *Int. Workshop on DYnamic languages for RObotic and Sensors systems (DYROS/SIMPAR)*, Germany, pp. 274–277.

Steck, A. & Schlegel, C. (2011). Managing execution variants in task coordination by exploiting design-time models at run-time, *Proc. IEEE/RSJ International Conference on Intelligent Robots and Systems (IROS)*, San Francisco, CA.

Szyperski, C. (2002). *Component-Software: Beyond Object-Oriented Programming*, Addison-Wesley Professional, ISBN 0-201-74572-0, Boston.

Tarr, P., Harrison, W., Finkelstein, A., Nuseibeh, B. & Perry, D. (eds) (2000). *Proc. of the Workshop on Multi-Dimensional Separation of Concerns in Software Engineering (ICSE 2000)*, Limerick, Ireland.

Vlissides, J. (2009). Pattern Hatching – Generation Gap Pattern.
URL: *http://researchweb.watson.ibm.com/designpatterns/pubs/gg.html*

Völter, M. (2006). MDSD Benefits - Technical and Economical.
URL: *http://www.voelter.de/data/presentations/mdsd-tutorial/02_Benefits.pdf*

Webber, D. L. & Gomaa, H. (2004). Modeling variability in software product lines with the variation point model, *Science of Computer Programming - Software Variability Management* 53(3): 305–331.

Willow Garage (2011). PR2: Robot platform for experimentation and innovation.
URL: *http://www.willowgarage.com/pages/pr2/overview*

# 5

# Programming of Intelligent Service Robots with the Process Model "FRIEND::Process" and Configurable Task-Knowledge

Oliver Prenzel[1], Uwe Lange[2], Henning Kampe[2],
Christian Martens[1] and Axel Gräser[2]
*[1]Rheinmetall Defence Electronics,*
*[2]University of Bremen*
*Germany*

## 1. Introduction

In Alex Proyas's science fiction movie "I, Robot" (2004) a detective suspects a robot as murderer. This robot is a representative of a new generation of personal assistants that help and entertain people during daily life activities. In opposition to the public opinion the detective proclaimed that the robot is able to follow his own will and is not forced to Isaac Asimov's three main rules of robotics (Asimov, 1991). In the end this assumption turned out to be the truth.

Even though the technological part of this story is still far beyond realization, the idea of a personal robotic assistant is still requested. Experts predicted robotic solutions to be ready to break through in domestic and other non-industrial domains (Engelberger, 1989) within the next years. But up to now, only rather simple robotic assistants like lawn mowers and vacuum cleaners are available on the market. As stated in (Gräfe & Bischoff, 2003), all these systems have in common that they only show traces of intelligence and are specialists, designed for mostly a particular task. Robots being able to solve more complex tasks have not yet left the prototypical status. This is due to the large number of scientific and technical challenges that have to be coped with in the domain of robots acting and interacting in human environments (Kemp et al., 2007).

The focus of this paper is to describe a tool based process model, called the "FRIEND::Process"[1], which supports the development of intelligent robots in the domain of personal assistants. The paper concentrates on the interaction and close relation between the FRIEND::Process and configurable task-knowledge, the so called process-structures. Process-structures are embedded in different layers of abstraction within the layered control architecture MASSiVE[2] (Martens et al., 2007). Even though the usage of layered control architectures for service robots is not a novel idea and has been proposed earlier (Schlegel &

---

[1] The name FRIEND::Process is related to the FRIEND projects (Martens et al., 2007). It has been developed within the scope of these projects, but is also applicable to other service robots.
[2] MASSiVE – Multilayer Control Architecture for Semi-Autonomous Service Robots with Verified Task Execution

Woerz, 1999; Schreckenghost et al., 1998; Simmons & Apfelbaum, 1998), MASSiVE is tailored for process-structures and thus is the vehicle for the realization of verified intelligent task execution for service robots, as it is shown in the following. The advantages of using process-structures shall be anticipated here:

- **Determinism**: Process-structures represent the complete finite sequence of actions that have to be carried out during the execution of a task. Due to the possibility of a bijective transformation from process-structures to Petri-Nets, a-priori verification with respect to deadlocks, reachability and liveness becomes possible. Thus, the task planner and executor, as part of the layered architecture, operate deterministically when using verified task-knowledge.

- **Real-time capability**: Additionally, the complexity of the task planning process satisfies real-time execution requirements, because this process is reduced to a graph search problem within the state-graph of the associated Petri-Net.

- **Fault-Tolerance**: Erroneous execution results are explicitly modeled within process-structures. Additionally, redundant behavior is programmatically foreseen. If an alternative robotic operation, which shall cope with the unexpected result, is not available, the user is included as part of a semi-autonomous task execution process.

To be able to provide a user-friendly configuration of process-structures and to guarantee consistency throughout all abstraction levels of task-knowledge, a tool-based process model – the FRIEND::Process – has been developing. The process model, on the one hand, guides the **development and programming** of intelligent behavior for service robots with process-structures. On the other hand, process-structures can be seen as a **process model for the service robot** itself, which guides the task execution of the robot during runtime. The unique feature of the FRIEND::Process in comparison to other frameworks (Gostai, 2011; Microsoft, 2011; Quigley et al., 2009) and the above mentioned control architectures is to completely rely on configurable process-structures and thus on determinism, real-time capability and fault tolerance.

The FRIEND::Process consists of the following development steps:

- **Analysis of Scenario and Task Sequence**: A scenario is split up into a sequence of tasks.

- **Configuration of Object Templates and Abstract Process-Structures**: The task participating objects are specified as Object Templates and pictographic process-structures on the symbolic (abstract) level are configured and verified.

- **Configuration of Elementary Process-Structures**: Process-structures on the level of system resources and sub-symbolic (geometric) information are configured and verified with the help of function block networks.

- **Configuration and Testing of Reactive Process-Structures**: Process-structures on the level of algorithms and closed loop control, operating sensors and actuators, are configured and tested, also with configurable function blocks.

- **Task Testing**: Task planning and execution is applied on all levels of process-structures and a complete and complex task execution is tested.

In the following Section 2, the motivation for the introduction of process-structures is explained in more detail by discussing the complexity of task planning for service robots with the help of an exemplary scenario. The description of the FRIEND::Process development steps is subject of Section 3. Throughout this description, exemplary process-structures of the sample scenario of Section 2 are introduced for each development step.

Finally, Section 4 summarizes and concludes the description of the FRIEND::Process for programming intelligent service robots.

## 2. Task planning on basis of process-structures

In this section, the complexity of classical task planning approaches is discussed first, before the introduction of process-structures is motivated. The discussion is carried out with the help of task execution examples from the field of rehabilitation robotics and the rehabilitation robot FRIEND III (IAT, 2009; Martens et al., 2007).

### 2.1 The complexity of classical task-planning approaches

With respect to one exemplary task – a service robot is supporting the preparation and the eating of a meal by a disabled person – the complexity of robotic task execution shall be illustrated. For this purpose the figures Fig. 1 to Fig. 3 are introduced. Fig. 1 shows the rehabilitation robot FRIEND III which is used as exemplary target system. In Fig. 2 snapshots of the task sequence "Meal preparation and eating assistance" are depicted. Finally, Fig. 3 shows the decomposition of this task sequence according to the principles to be presented in detail in this paper.

FRIEND III is a general purpose semi-autonomous rehabilitation robot suitable for the implementation of a wide range of support tasks. As depicted, FRIEND III consists of an electrical wheelchair which is equipped with several sensors and actuators: A stereo camera system mounted on a pan-tilt-head, force torque sensor, robotic arm and gripper with force control. FRIEND III has been developed by an interdisciplinary team of engineers, therapists and designers and has been tested with disabled users within the AMaRob project (IAT, 2009).

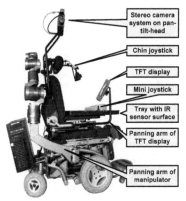

Stereo camera system on pan-tilt-head

Chin joystick

TFT display

Mini joystick

Tray with IR sensor surface

Panning arm of TFT display

Panning arm of manipulator

Fig. 1. FRIEND III rehabilitation robot

To perform "meal preparation and eating assistance", the robot system has to execute the following actions:

- Locate the refrigerator, open the refrigerator door, locate the meal inside the refrigerator, grasp and retrieve the meal from the refrigerator, close the refrigerator door
- Open the microwave-oven, insert the meal, close the oven, start the heating process

- Open the microwave-oven door again, grasp and retrieve the meal, close the microwave-oven door
- Place the meal in front of the user, take away the lid
- In a cycle, take food with the spoon and serve it near the user's mouth, finally put the spoon back to the meal-tray
- Clear the wheelchair tray

Fig. 2. Task sequence for meal preparation and eating assistance

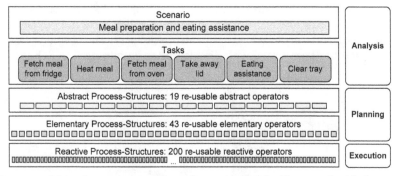

Fig. 3. Decomposition of a scenario on four abstraction levels, illustrated with the sample scenario "Meal preparation and eating assistance"

As shown in Fig. 3, the overall scenario is decomposed into tasks, abstract operators, elementary operators and reactive operators according to the layered control architecture MASSiVE. Abstract process-structures ($PS_A$[3]) model behavior on task planning level and elementary process-structures ($PS_E$) model behavior on system planning level. The reactive process-structures ($PS_R$) define reactive operations on the executable algorithmic level. From viewpoint of task planning, the "meal preparation and eating assistance" scenario is split up into 6 tasks, 19 abstract operators and 43 elementary task planning operators. Additionally, a large set of reactive operators is required within the execution layer.

In typical human environments, it is impossible to predefine a static sequence of operators beforehand. Many dynamic aspects resulting from dynamic environmental changes have to be

---

[3] Find all abbreviations in the glossary at the end of this paper.

considered, e. g. caused by changing lighting conditions, arbitrarily placed and filled objects, changing locations of objects and the robotic platform, various obstacles, and many more. Consequently, a strategy to plan a sequence of actions that fulfills a certain task is mandatory. Many task planners are based upon deliberative approaches according to classical artificial intelligence. Typically, the robotic system models the world with the help of symbolic facts (e. g. first order predicate logic, (Russel & Norvig, 2003)), where each node of a graph represents a state (snapshot) of the world. The planner has to find a sequence of operations which transforms a given initial state into a desired target state. In the worst cases this leads to NP-complete problems, as there is an exponential complexity of classical search algorithms (Russel & Norvig, 2003). If we consider breadth-first search as a simple example, a calculation time of hours results at search depth 8; and with a depth of 14, hundreds of years are required for exhaustive search (branching factor 10 and calculation time of 10.000 nodes/s are assumed). The search depth is related to the number of required operators for a certain task and the branching factor results from the number of applicable operators in one node. Compared to the number of required and available operations shown in Fig. 3 it becomes obvious that only trivial problems can be solved on this basis. Certainly, the mean search time can be improved in comparison to breadth-first search, with e. g. heuristic approaches like A*, with hierarchical planning, search in the space of plans or successive reduction of abstraction (Russel & Norvig, 2003; Weld, 1999), but in worst cases a planning complexity as mentioned has to be faced. Even though the improvements of deliberative task planners are notable, it is still questionable whether they are efficient (real-time capable) and robust (deterministic and fault-tolerant) enough for the application in real world domains (Cao & Sanderson, 1998; Dario et al., 2004; Russel & Norvig, 2003).

## 2.2 Process-structures as alternative to classical planning approaches

An alternative to deliberative systems are assembly planning systems. Cao and Sanderson proposed such an approach for the application to service robotics (Cao & Sanderson, 1998). Based on this idea, Martens developed a software-technical framework (Martens, 2003) that operates on pre-structured task-knowledge, called *process-structures*. Table 1 summarizes the concept of process-structures and the distinction of task level, system level and algorithmic level.

Fig. 4 shows an example of an abstract process-structure that models the fetching of a cup from a container. The object constellations (OC) model the physical contact situation of the involved objects box (B), container (C), gripper (G) and table (T). The object constellations are connected via composed operators (COPs). These are in most cases (i. e. where this is physically meaningful) bi-directional operators. To be able to perform task planning based on an abstract process-structure, a set of OCs defines an initial situation and another set of OCs defines the target situation. Thus, task planning on abstract level means to find a sequence of COPs from initial to target situation. The initial situation is usually dynamically determined at runtime with the help of an initial monitoring procedure (Prenzel, 2005). The target situation is pre-determined for a certain $PS_A$.

A process-structure contains a context-related subset of task-knowledge. The finite size of a process-structure makes planning in real-time with short time intervals as well as a priori verification possible. The logical correctness of a structure is checked against a set of rules. A positive result of this check guarantees that no system resource conflicts exist. It also guarantees the correct control and data flow. Altogether, the concept of process-structures is the basis for a robust system runtime behavior. Despite pre-structuring, the process-structures are still flexible to adapt to diverse objects, so that their re-usability in different scenarios is achieved. Technical details of process-structures beyond this summarized concept description can be found in (Martens et al., 2007).

| $PS_A$ | Task Level |
|---|---|
| Defines what happens | Models e. g. the fetching of an object |
| Is configured by: | Non-technical personnel or the user |
| $PS_E$ | System Level |
| Defines how something happens from system perspective | Models the usage of system resources and the control and data flow |
| Is configured by: | System programmer |
| $PS_R$ | Algorithmic Level |
| Defines how something happens from perspective of reactive algorithms | Models the combined usage of hardware sensors and actuators |
| Is configured by: | System programmer |

Table 1. Summarized concept of process-structures

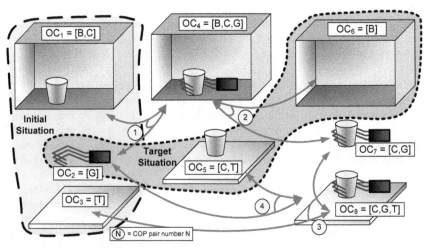

Fig. 4. Schematic illustration of an abstract process-structure ($PS_A$) which models the fetching of a cup from a container-like place as e. g. a fridge or a cupboard

The applicability of process-structures for the programming of service robots has been shown in (Martens, 2003) with the help of several representative rehabilitation robotic scenarios. As anticipated in the introduction this approach has been extended during the AMaRob project (2006 – 2009) and within (Prenzel, 2009) to embed the process-structure-based programming into a process model – the FRIEND::Process. From task analysis to final testing of implemented system capabilities, the FRIEND::Process guides through the complete development cycle of a service robot based on a closed chain of user-friendly configuration tools. Enhancements of the FRIEND::Process are matter of ongoing developments.

## 3. The FRIEND::Process

Process models structure complex processes in manifold application areas. With respect to
system- and software-engineering, a process model shall organize the steps of development,
the tools to be used and finally the artifacts to be produced throughout the different
development stages. The overall scheme of the FRIEND::Process is depicted in Fig. 5.
Central elements of the process and consequently the specialty in comparison to other
process models are the process-structures. Within the development steps, the building
blocks of process-structures are decomposed as shown in Table 2. In the following sections
the five development steps of the FRIEND::Process are discussed in detail. Thus, the
contents of Table 2, i. e. the composition of process-structures and the decomposition on the
next level as well as the abbreviations will be explained. Also, the application of the
FRIEND::Process for the development of the sample task of "meal preparation and eating
assistance" is shown in each step.

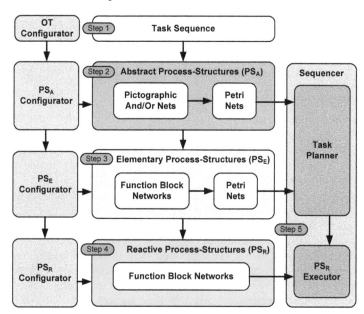

Fig. 5. Scheme of the FRIEND::Process with five development steps and the respective
process-structure levels as well as the involved tools for configuration, planning and
execution

| Process-Structure Decomposition | Process-Structure Building Blocks |
|---|---|
| Scenario → Task Sequence | Tasks |
| Task → PS$_A$ | System, Object Templates (OTs), Object Constellations (OCs), Facts, Composed Operators (COPs) |
| COP → PS$_E$ | System, Object Templates (OTs), Facts, Skills |
| Skill → PS$_R$ | System, Object Templates (OTs), Reactive Blocks |

Table 2. Decomposition and building blocks of process-structures

### 3.1 FRIEND::Process step 1: Analysis of scenario and task sequence

Development according to the FRIEND::Process starts with the "Scenario Analysis" as step 1. Unlike the subsequent steps, this step is not (yet) tool-supported. The scenario analysis splits up a complex scenario like "meal preparation and eating assistance" into a sequence of re-usable tasks. Also, a structured collection of the objects takes place that are in the focus of a certain scenario.

### 3.1.1 Description of the process step

The development step 1 is dedicated to a first analysis of the desired task execution scenario. As shown in Fig. 6 a sequence of re-usable tasks is specified. Besides the strictly sequential concatenation of tasks, cyclic repetitions are also possible, as e. g. required for the eating assistance scenario introduced at the beginning of the paper.

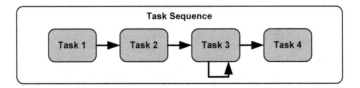

Fig. 6. A complex task sequence consists of several tasks

The FRIEND::Process defines criteria for task splitting:

- **Modularity, low complexity and re-usability:** One task is focusing on a set of objects. This set shall be kept as small as possible to limit the task's complexity and to ensure re-usability of a task. It shall be possible to use the tasks independently, but also to concatenate them to more complex action sequences.
- **The typical physical location of the objects:** If movement of the robotic platform is required, this is a clear indicator to switch the task context, e. g. when moving from fridge to microwave oven in the meal preparation scenario. After moving the platform, relative locations between platform and objects have to be re-assessed.

Currently, the process step 1 is not yet supported by a dedicated tool. Therefore, to still achieve a certain level of formality, the results of scenario analysis are collected in a UML use case diagram as seen in Fig. 7. For each task a use case with verbal task description is specified. This includes the objects involved in the task, the so-called task participating objects (TPO).

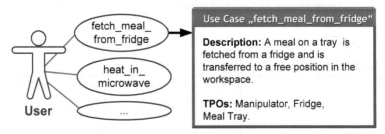

Fig. 7. Use case diagram with tasks (use cases) of the sample scenario. For each task, a detailed description as well as the set of task participating objects (TPO) is specified

The objects involved in task execution are the elements that are relevant in all subsequent development steps. To follow the principle of re-usable task-knowledge, the TPOs are specified as abstract object classes. For example, a task that describes the fetching of an object from a container-like place (see Fig. 4) can be re-used to fetch either a bottle or a meal from the refrigerator. In the FRIEND::Process the re-usable classes of objects are specified as hierarchical UML ontology. An exemplary ontology for the scenario "meal preparation and assistance to eat" is depicted in Fig. 8. It is depicted that the TPOs are constructed from basic geometric bodies (cuboid and cylinder) and more complex objects are created with inheritance and aggregation.

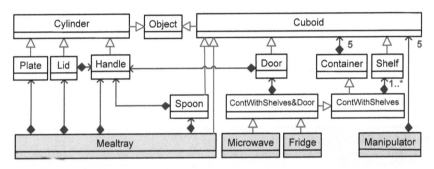

Fig. 8. Ontology of task participating objects (TPO) for the scenario "Meal preparation and assistance to eat"

To embed the TPOs in the tool-chain that covers all further development steps, the concept of "Object Templates" (OT) has been introduced (Kampe & Gräser, 2010). The configuration of Object Templates and their integration into the different levels of process-structure configuration will be discussed in more detail within the following process steps.

### 3.2 FRIEND::Process step 2: Configuration of object templates and abstract process-structures

In this development step the task participating objects are formally specified and configured with the help of Object Templates. Subsequently, an abstract process-structure (PS$_A$) is configured based on pictographic And/Or-Nets. This means that physical object constellations (OC) and physical transitions between the object constellations are specified. Besides configuration of PS$_A$, the logical correctness of the abstract process-structures is guaranteed by the configuration tool. Finally, the pictographic PS$_A$ are converted to Petri-Nets according to (Cao & Sanderson, 1998) for the input into the task planner.

In the following, a description of the process step is introduced first. Afterwards, the configuration concept for Object Templates is shown. Finally, the configuration of an abstract process-structure is exemplified.

### 3.2.1 Description of the process step

As shown in Fig. 9 the FRIEND::Process decomposes each task into an abstract process-structure (PS$_A$). A schematic exemplary pictographic PS$_A$ for the task "Fetch cup from container" has already been introduced and discussed in Fig. 4. Within the FRIEND::Process, the configuration of PS$_A$ is carried out within a pictographic configuration environment, the so-called $PS_A$-Configurator. Fig. 10 shows the $PS_A$-Configurator with the $PS_A$ "Fetch meal from fridge".

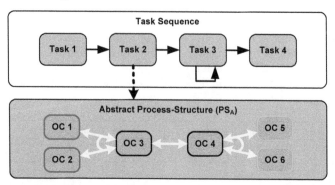

Fig. 9. Decomposition of a task as abstract process-structure with object constellations (OC) and composed operators (COP)

The procedure of $PS_A$ configuration is as follows:
- Selection of task participating objects (TPOs)
- Composition of object constellations (OCs)
- Connection of OCs via composed operators (COPs)
- Selection of default initial and default target situation

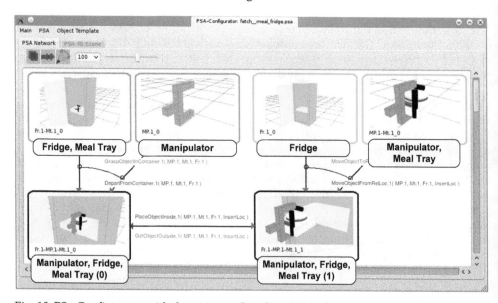

Fig. 10. $PS_A$-Configurator with the pictographic abstract process-structure modeling the Task "Fetch meal from fridge"[4]

---

[4] For better readability, overlays have been added in this illustration.

The pictographic representation of an OC is configured within a sub-dialog within the $PS_A$-Configurator. Within this configuration dialog, the predicate logic facts, which are assigned to an OC, can be inspected. These facts are the pre- and post-condition facts of the COPs that interconnect the OCs. Within the constraints given by the COP facts, the pictographic appearance of an OC can be adjusted within the $PS_A$-Configurator within a 3D scene. The rendering of object constellations is based on "Object Templates".

### 3.2.2 Object templates

Objects play a central role in process-structures. The different levels of process-structures model different aspects of objects. On abstract level, a symbol is associated with an object for the purpose of task planning (e. g. "Mt.1" for the meal tray in the sample scenario). On system level, i. e. on the level of elementary process-structures, so-called sub-symbolic (i. e. geometric) object information is processed. With respect to the meal tray this is, for instance, the location to grasp the tray. To model the different aspects of objects and to assure an information consistency throughout the different information layers, the concept of Object Templates has been introduced.

Object Templates comprise the following aspects:

- A 3D model of the object, used for pictographic rendering of object constellations on $PS_A$ level as well as for motion planning and collision avoidance on $PS_R$ level
- Associated sub-symbolic (geometric) data for planning and execution on $PS_E$ and $PS_R$ level, e. g. the grasping location of an object
- Complex objects can be composed of simpler objects; e. g. a meal tray consists of a tray, a plate, a lid and a spoon
- Object Templates are configured with natural parameters of the composed object, e. g. width, height, depth and wall thickness for a container, instead of separate specification of all geometric primitives
- The 3D appearance of Object Templates is associated with task-knowledge like symbolic facts and characteristics. For example the fact "IsAccessible(MicrowaveOven)" renders the opening status of the door of the oven's 3D model.

An exemplary Object Template is the meal tray depicted in Fig. 11. It consists of a base tray, a plate with a lid and a spoon. Both the lid and the spoon are detachable from the meal tray. The different stages of separation are depicted in Fig. 12.

Fig. 11. The meal tray of the eating scenario as photo (left) and modeled by means of an Object Template (right)

Fig. 12. The different separation stages (detached lid, detached spoon, both lid and spoon detached) of the meal tray

The configuration of Object Templates takes place within the Object-Template-Configurator (OT-Configurator) which is part of the $PS_A$-Configurator as shown in Fig. 13.

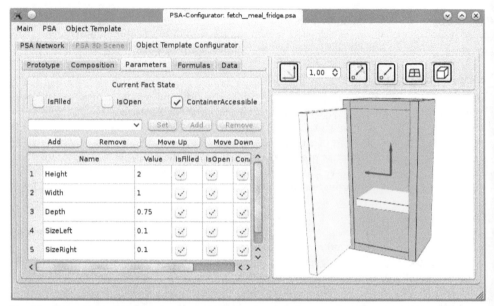

Fig. 13. Object-Template-Configurator (OT-Configurator) as part of the $PS_A$-Configurator

Within the screenshot in Fig. 13 the Object Template of the refrigerator is modeled. On the left side the parameters and their association to symbolic facts are specified. On the right side the 3D model of the object is rendered according to the current configuration. To render the 3D model of a composed object, the aggregated sub-objects are composed with formulas within the Object-Template-Configurator tool. Frequently required and complex formulas like alignment and rotation of Object Templates are provided with the help of assistive functions.

Certain aspects of the 3D geometry have a fix association with object characteristics as given in the following table:

| Characteristic | Associated sub-symbolic element |
|---|---|
| IsGrippable | Coordinates to grasp the object |
| IsPlatform | Limits to place other objects onto this object |
| IsContainer | Limits to place other objects within this object |

Table 3. Relations between characteristics and sub-symbolic elements

### 3.2.3 Exemplary abstract Process-Structure: Fetch meal tray from fridge

The exemplary $PS_A$ that shall be discussed in detail has already been introduced within the $PS_A$-Configurator frontend in Fig. 10. In this $PS_A$ the task participating object are a fridge (symbol "Fr" with instance number "1" → "Fr.1"), a meal tray ("Mt.1"), the manipulator ("MP.1") and the abstract symbol for a relative location ("InsertLoc"). In this $PS_A$ the initial situation consists of two object constellations. The first one models the manipulator in a free position in the workspace (instance number "0" is assigned to this object constellation → "MP.1_0"). The second object constellation models the already opened fridge containing the meal tray ("Fr.1-Mt.1_0"). The two OCs are connected via the assembly COP "GraspObjectInContainer(MP.1, Mt.1, Fr.1)". If physically possible, a complementary disassembly operator is assigned to model the reverse operation for re-usage of the PS$_A$ in another scenario context. In this case this is the COP "DepartFromContainer(MP.1, Mt.1, Fr.1)". The assembled object constellation is depicted on the bottom left side and the associated abstract planning symbol is "Fr.1-MP.1-Mt.1_0". Due to the associated symbolic facts, which are imposed within the object constellation by the post-condition facts of the COP, the pictographic representation is rendered so that the manipulator grasps the meal tray in the fridge.

Besides assembly and disassembly operators, the And/Or-Net syntax provides operators modeling the internal state transition (IST) of object constellations (IST COPs). IST COPs are applied when the physical contact state of the involved objects is not changed. From the viewpoint of planning on abstract level, objects being in close relative locations to each other are considered to be in a physical contact situation. Therefore, the IST COP "GetObjectOutside(MP.1, Mt.1, Fr.1, InsertLoc)" is applied to transform the OC "Fr.1-MP.1-Mt.1_0" on the left side into the OC "Fr.1-MP.1-Mt.1_1" on the right side. Finally, the COP "MoveObjectFromRelLoc(MP.1, Mt.1, Fr.1, InsertLoc)" models the disassembly operation and results in two object constellations which model the target situation of this abstract process-structure: "Fr.1_0" is the empty fridge and "MP.1-Mt.1_0" is the manipulator with the gripped meal tray in a free position in the work space.

To be able to develop and verify the three levels of process-structures independently, i. e. in a modular manner, the consistency of task-knowledge on all levels has to be assured. This is achieved with common building blocks of the different process-structures as shown in the decomposition chain in Table 2. The common elements are the interfaces to the next level of process-structures. The important interfacing elements between PS$_A$ and PS$_E$ are the pre- and post-condition facts of the COP to be decomposed as PS$_E$ in the next process step. For the COP "GraspObjectInContainer" the facts are shown in Table 4.

| Pre-Facts | Post-Facts |
|---|---|
| HoldsNothing(Manipulator) = True | HoldsNothing(Manipulator) = False |
| IsInFreePos(Manipulator) = True | IsInFreePos(Manipulator) = False |
| - | IsGripped(Manipulator, Object) = True |
| ContainerAccessible(Container) = True | - |
| IsInsideContainer(Object, Container) = True | - |

Table 4. Pre- and Post facts of COP "GraspObjectInContainer(Manipulator, Object, Container)"

### 3.3 FRIEND::Process step 3: Configuration of elementary process-structures

In the third process step, each composed operator (COP) of an abstract process-structure (PS$_A$) is decomposed into an elementary process-structure (PS$_E$). To achieve user-friendly configuration of PS$_E$, configurable function blocks are assembled to function block networks (FBN). Each function block models a reactive robot system operation, also called skill. A priori verification of task-knowledge on this level takes place with the help of Petri-Nets, which result from automatic conversion of FBNs.

### 3.3.1 Description of the process step

Fig. 14 depicts the decomposition principle of COPs into elementary process-structures, consisting of skill blocks. An elementary process-structure, as first introduced by (Martens, 2003), is a Petri-Net with enhanced syntax and superordinated construction rules. The advantage of Petri-Nets is their ability to model parallel activities. This is useful for the behavioral modeling on robot system level, for instance, if a manipulator action is guided by input from a camera system or another sensor. Furthermore, Petri-Net-based $PS_E$ offer mathematical methods for analysis of the reachability of a certain system state, for verification of the correctness of control and dataflow and for the exclusion of resource conflicts (Martens, 2003).

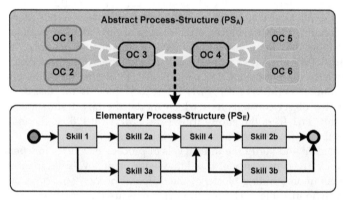

Fig. 14. Decomposition of a composed operator (COP) as elementary process-structure

Besides these conceptual advantages, from the viewpoint of implementation it turned out that the programming of elementary process-structures with Petri-Nets is a time consuming and error prone procedure. The setup of a correctly verified Petri-net-$PS_E$ usually takes several hours. Even with strong modularization of the networks, the large number of places and transitions leads to hardly manageable Petri-Nets in real-life applications. This is the reason why the FRIEND::Process introduces the configuration of $PS_E$ on the basis of function block networks (FBN). Similar to the $PS_A$-Configurator, a configuration frontend, called $PS_E$-Configurator, has been created. This tool subsumes all logical and syntactical rules that are required for $PS_E$-configuration. Furthermore, a conversion algorithm has been developed (Prenzel et al., 2008), which converts an FBN into a Petri-Net for automatic execution of verification routines, like a reachability analysis. A screenshot of the $PS_E$-Configurator with the $PS_E$ "GraspObjectInContainer" is given in Fig. 15.

With respect to their representative function for Petri-Nets, the control flow within the FBN structures is token-oriented. The execution starts from the "Start" block and ends at the "Target Success" block. In-between, reactive skills are executed, including manipulative operations as well as sensor operations or user interactions. Each function block has one input port, and several output ports according to the possible execution results of the skill (see e. g. block "CoarseApproachToObjectInContainer" in Fig. 15 with the output ports "Success", "Failure", "Abort" and "UserTakeOver"). The output port "Abort" is not explicitly connected to an abort block to increase the readability of the network structure. The typical construction rule for a semi-autonomous system (like FRIEND) is to provide user interactions as redundant action for autonomous system operations. As shown in Fig. 15, the failure of an autonomous operation (e. g. "AcquireObjectBySCam") is linked to the user interaction "DetermineObjectBySCam", replacing the failed system action.

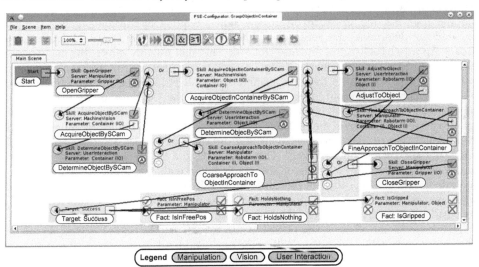

Fig. 15. PS$_E$-Configurator with elementary process-structure as function block network, modeling the COP "GraspObjectInContainer".[5]

The configuration of PS$_E$ on the basis of function block networks does not only achieve a decisive increase of development comfort (configuration instead of programming), but it also decreases the required task-knowledge engineering time significantly. By building the PS$_E$ directly in the correct manner, the time-consumption for the construction of one PS$_E$ is reduced from hours to 10-15 minutes per network. On this basis, real world problems like the "Meal preparation and assistance" task, become manageable in their complexity.

### 3.3.2 Exemplary elementary process-structure: Manipulator grasps meal tray in fridge

The exemplary PS$_E$ "GraspObjectInContainer", as shown in Fig. 15, models the grasping of an object in a container-like place in a general way. In the sample scenario "meal preparation" this PS$_E$ is applied to fetch the meal tray from the refrigerator and also from the microwave oven after heating of the meal.

The objects (Object Templates) "Manipulator", "Object" and "Container", which are involved in this PS$_E$, are the input artifacts handed over as COP parameters from the previous step of the FRIEND::Process. The first skill block that follows the "Start" block is the manipulator skill "OpenGripper". Subsequently, the container (fridge) is located with the help of the vision skill "AcquireObjectBySCam(Object)". This skill calculates the location and size of the given object with the help of a stereo camera (SCam). In the sample scenario the COP parameter "Container" (i. e. the fridge) is inserted at the skill's placeholder

---

[5] For better readability, overlays have been added in this illustration.

"Object" according to the principle of type-conform parameter replacement (Martens, 2003). The Object Template of a fridge provides the according two sub-symbolic parameters location and size. A successful execution of the skill guarantees that the container's location and size are stored in the system's world model and can serve as input parameters for subsequent skills. After verification of the associated Petri-Net of this PS$_E$ the correctness of the data flow between all skill blocks is assured. If the recognition of the fridge is successful and the user has not to be involved, the skill "AcquireObjectInContainerBySCam" is executed to determine the location of the meal tray in the fridge. Afterwards, a "CoarseApproachToObjectInContainer" follows. This skill roughly directs the manipulator in front of the meal tray in the fridge based on the location information calculated beforehand. Fig. 15 depicts that this manipulator skill is followed by an enforced user interaction, since all output ports are connected to the Or-block preceding the user interaction. The confirmation by the user is included at this place due to testing purposes to assure a correct execution of the first skill. For real task-execution a quick reconfiguration of the PS$_E$ will change the system behavior and directly execute the next manipulator skill "FineApproachToObjectInContainer". This skill leads to a final grasping of the meal tray handle, while avoiding collisions of the manipulator with the fridge with the help of dedicated methods for collision avoidance and path planning (Ojdanic, 2009). The final action necessary to complete the grasping is to close the gripper. The PS$_E$ ends with setting the post-facts of the COP as specified in Table 4.

From the viewpoint of the system's task planner, each skill-function-block represents an elementary (executable) operation. Within the execution level of the system, the operations are not seen as atomic units. The execution of one skill means to activate reactive system functionality, for instance the sensor-controlled approach of an object to be grasped in the skill "FineApproachToObjectInContainer". These basic system skills have to couple sensors and actuators on the algorithmic level. To pursue the paradigm of configurable process-structures also on this level, the FRIEND::Process introduces reactive process-structures.

### 3.4 FRIEND::Process step 4: Configuration and testing of reactive process-structures

Historically, during the elaboration of the FRIEND::Process, the elementary operators (skills) have been implemented directly in C++. Subsequently, when appropriate CASE-tools became available, the elementary operators have been implemented with model driven development techniques (Schmidt, 2006) as executable UML models. Then, a configuration tool has been developed, which makes user-friendly configuration of process-structures possible also on this development level. With the help of this tool it is assured that the verified interfaces from the PS$_E$-layer are respected and the robustness assertion throughout the complete system architecture is maintained.

### 3.4.1 Description of the process step

Fig. 16 depicts the decomposition of a skill block from PS$_E$-layer into a reactive process-structure (PS$_R$) consisting of algorithmic blocks. Similar to the PS$_E$ function blocks, PS$_R$ are also based on configurable function block networks. The PS$_R$-Configurator tool results from the Open-Source Image Nets Framework[6], which originally has been developed for configurable image processing algorithms.

---

[6] http://imagenets.sourceforge.net/

The $PS_R$ Configuration Framework consists of the following five parts (see Fig. 17):

- $PS_R$-Configurator,
- Embedding of $PS_R$ into any C++ code via $PS_R$-Executor,
- Reactive process-structures ($PS_R$), which are executable function block networks,
- Extensible set of Plug-Ins and
- Configurable function blocks

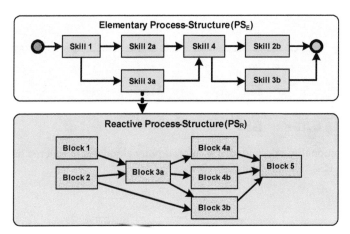

Fig. 16. Decomposition of a skill into a reactive process-structure

The "PSR-Configurator" is a graphical user interface, which can be used to rapidly create a "function block network", namely a reactive process-structure ($PS_R$). With the $PS_R$-Executor, it is possible to load and execute the previously configured $PS_R$. The $PS_R$ itself is a directed graph, connecting configurable "function blocks". Each block can execute code to process its input (image data or other data) and save its outputs. One or more blocks are grouped in a "Plug-In" and an arbitrary number of Plug-Ins can be loaded dynamically by the $PS_R$. In this way, the $PS_R$-Framework can be easily extended by new independent Plug-Ins. This independency of the algorithmic modules results in completely independent development within a team of developers. In addition, the strong modularization leads to a technically manageable amount of code within a single block and reduces the time of inspecting an erroneous block. The $PS_R$ execution library can save a $PS_R$ in human readable XML format. Thus, on the one hand the $PS_R$-Configurator can configure, load and save a $PS_R$, but on the other hand also external C++ code can load a $PS_R$ file.

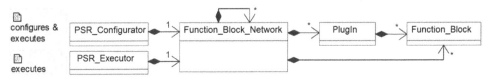

Fig. 17. The UML structure of the $PS_R$-Configuration Framework

Hierarchical modeling is a common method to subdivide algorithms into separate parts - it breaks down the complexity and facilitates reusability. In the $PS_R$-Framework, parts can be constructed as separate $PS_R$ and can be combined afterwards to constitute a complete algorithm. The $PS_R$-Executor is in fact also a function block, which can load and process a $PS_R$. The connection between the $PS_R$ inside an Executor and the outer net is established by special input and output blocks. For example the $PS_R$ "Color2Color3D" shown in Fig. 18 calculates a colored point cloud out of a stereo image pair. On the left side there are two input blocks, which hand over the images from the block in orange. This block only exists in this $PS_R$ for testing the net and will be ignored on execution if this $PS_R$ is loaded by a $PS_R$-Executor (see Fig. 19, right side).

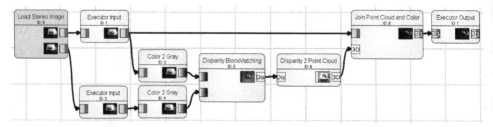

Fig. 18. The functionality of calculating a colored point cloud out of a stereo image pair is depicted in this $PS_R$, called "Color2Color3D"

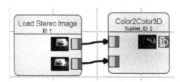

Fig. 19. The previously shown $PS_R$ can be loaded as one $PS_R$-Executor block

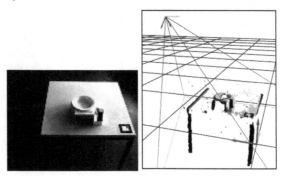

Fig. 20. Left: original image, right: resulting point cloud of the stereo camera images visualized in 3D by the $PS_R$-Configurator

The $PS_R$ in Fig. 19 shows the use of a subnet of an image acquisition together with the calculation of the extrinsic matrices of a stereo camera, which describe the relation of the cameras to the robot. These matrices depend on an invariant transformation frame inside the pan-tilt-head (see Fig. 1) and its rotation angles. By combining the two subnets, a live view of the stereo camera's point cloud can be calculated (depicted in Fig. 20, in the center of the images the meal tray can be seen). As a visually guided robot is a real world object, which moves in the three dimensional Cartesian space, it is useful to display the vision results in the same space. While configuring a PS$_R$ with the PS$_R$-Configurator, intermediate results can be visualized in two and three dimensions; depending on the data type, for example scalar values can only be visualized in 2D, camera matrices can be visualized in 2D and 3D (using OpenGL (Wright et al., 2010), see Fig. 20, right).

To be able to execute a PS$_R$ as skill block within the context of the PS$_E$ layer and to guarantee that the PS$_E$ interfaces are respected, a special type of "Verified PS$_R$-Executor block" is created. During configuration of this kind of block, the PS$_R$-Configurator checks that the used resources as well as input and output parameters match the specification of a certain PS$_E$ skill to be modeled as PS$_R$. For example in the case of the PS$_R$ "AquireObjectBySCam(Object)" the allowed resource is the stereo camera system. The input parameter is the Object Template of the given object and the output parameter are the return values "Success" and "Failure".

### 3.4.2 Exemplary reactive process-structure: Acquire meal tray by stereo camera

To show the capabilities of the reactive process-structures, a simplified example is discussed in the following, namely the machine vision skill to acquire an object by the stereo camera with the configuration "Meal Tray". This example of a PS$_R$ is non-reactive, as no actor is involved. Though, in a more complex PS$_R$, it is possible to combine the camera and the robot in a feedback loop to implement visual servoing to achieve reactive behaviour.

In Fig. 21 several general blocks are used to find the red meal tray handle in an image. The processing chain starts with the detection of highly saturated, red parts. It is followed by a 9x9 closing operation to eliminate noise. Afterwards, contours are detected and filtered according to a priori knowledge of the size of the handle. Then, the minimum rectangles around the contours are determined and the major axes and their end points are calculated. For testing the current PS$_R$, again the orange blocks have been added to visualize intermediate testing results and they are not executed during task execution.

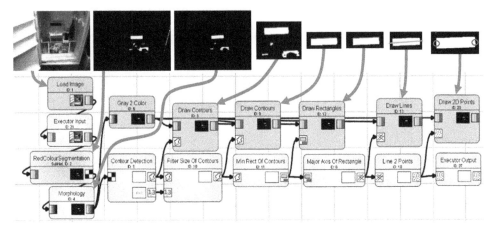

Fig. 21. PS$_R$ "MajorAxisPoints" which detects red areas of a certain size and calculates the major axes of these areas. Orange blocks are omitted when this PS$_R$ is used in a PS$_R$-Executor

To grasp the meal tray handle with the manipulator, the determination of its location in 3D is required. Thus, a 2D detection of the meal tray is not sufficient. However, the previously created and tested PS$_R$ "MajorAxisPoints" can be used twice, one for each image of the stereo camera. Fig. 22 depicts the usage of the previous net to calculate the 3D line, describing the handle of the meal tray. The block *Optimal Stereo Triangulation* computes a 3D contour based on key feature points, extracted from a stereo image. With the known camera matrices and the 2D feature correspondences, the 3D points are found by the intersection of two projection lines in the 3D space using optimal stereo triangulation, as described in (Natarajan et al., 2011).

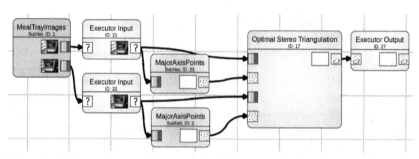

Fig. 22. $PS_R$ which detects the meal tray handle in 3D

Next, the 3D line of the 3D handle detection is used to calculate a transformation frame, having the position of the right 3D point and the rotations to point the y-axis in line direction. Using the a priori knowledge that the meal tray should be parallel to the world coordinate system, only rotation around z-axis has to be calculated. Fig. 23 displays (top, from left to right) the 3D line, the calculated frame, the meal tray Object Template and the placed meal tray, based on the frame. For the fulfillment of the specification of the calling $PS_E$, the Object Template has to be written to the World Model (a service to read from and write data to) with the "Write to World Model" block. This ensures that the detected object is globally available for later processing steps and is the actual result of this $PS_R$.

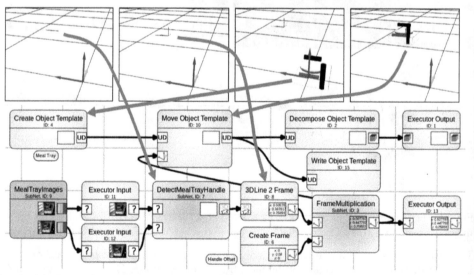

Fig. 23. Meal tray detection and Object Template placement based on 3D handle detection, frame calculation and Object Template movement (UD = user data)

For simulation and $PS_R$ unit testing in the $PS_R$-Configurator, the fridge, the static environment (wheelchair, monitor and user) and the robot with its current configuration can be placed in the same 3D scene with the meal tray. Fig. 24 and Fig. 25 show the real scene and the 3D simulation result in comparison.

Fig. 24. Real scene of this $PS_R$

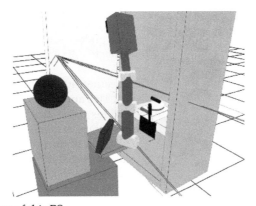

Fig. 25. Simulated scene of this $PS_R$

### 3.5 FRIEND::Process step 5: Task testing
After finishing the configuration of process-structures on all three levels, the planning and execution of a task ($PS_A$) has to be tested. The modularly configured, verified and tested process-structures of lower abstraction ($PS_E$ and $PS_R$) are involved in this final process step.

### 3.5.1 Description of the process step
For the purpose of task testing the "Sequencer" is used, which embeds a task planner for process-structures and the $PS_R$-Executor (see Fig. 5). The Sequencer is part of the process-structure-based control architecture MASSiVE mentioned in Section 1. The Sequencer interacts with skill servers, which offer the functionality that has been configured and

verified as reactive process-structures beforehand. The layered system architecture organizes a hardware abstraction via skill layer, so that there is a unique access-point on the sensors and actuators from a certain responsible skill server.

Task tests can be performed in the following execution modes:

- *Probabilistic simulation:* the skill interfaces and the communication infrastructure are tested and skill return values are simulated,
- *Skill simulation:* the skill's functional core is simulated,
- *Motion simulation:* the motion governed by manipulative skills is simulated and visualized within a virtual 3D space as shown in Fig. 25,
- *Hardware simulation:* the sensors and actuators are simulated,
- *Real execution:* the skill is executed with access of sensors and actuators.

Based on the process-structures, a complete task is planned and executed in one of the listed skill execution modes. This means that the Sequencer first plans a sequence of COPs and subsequently decomposes each COP into an elementary process-structure. Planning on this level results in a sequence of skills to be executed then. Step by step and based on the execution result of each skill, the once planned skill sequence is pursued, or re-planning takes place if an unexpected result is obtained.

## 4. Conclusion

As shown in Section 2.1 it is a challenging task to establish intelligent behavior of service robots operating in human environments. Typical operation sequences of support tasks in daily life activities seem to be simple from human understanding. However, to realize them with a robotic system, a huge complexity arises due to the variability and unpredictability of human environments.

In this paper the FRIEND::Process – an engineering approach for programming robust intelligent robotic behavior – has been presented. This approach is an alternative solution in contrast to other existing approaches, since it builds on configurable process-structures as central development elements. Process-structures comprise a finite-sized and context-related set of task knowledge. This allows a priori verification of the programmed system behavior and leads to deterministic, fault-tolerant and real-time capable robotic systems.

The FRIEND::Process organizes the different stages of development and leads to consistent development artifacts. This is achieved with the help of a tool chain for user-friendly configuration of process-structures.

The applicability of the here proposed methods has been proven throughout the realization of the AMaRob project (IAT, 2009) where task execution in three complex scenarios for the support of disabled persons in daily life activities has been solved. One of theses scenarios is the "Meal preparation and eating assistance" scenario, used for exemplification throughout this paper. The most error prone and thus challenging action in this scenario is the correct recognition of smaller objects (e. g. the handle of the meal tray) under extreme lighting conditions. However, with the inclusion of redundant skills in the elementary process-structures, the system's robustness has been raised in an evolutionary manner. In cases where even redundant autonomous skills did not execute successfully, the accomplishment of the desired task was achieved via inclusion of the user within a user interaction skill.

Currently, the methods and tools discussed in this paper are continuously developed further and are applied in the project ReIntegraRob (IAT, 2011). The mid-term objective is to integrate the different configuration tools for process-structures into one integrated configuration environment. The $PS_R$ Configuration Framework, which is the most elaborated tool, will build the basis for this.

## 5. Glossary

| COP | Composed Operator |
|-----|-------------------|
| FBN | Function Block Network |
| FRIEND | Functional Robotarm with user-frIENdly interface for Disabled people |
| MASSiVE | Multilayer Control Architecture for Semi-Autonomous Service Robots with Verified Task Execution |
| OC | Object Constellation |
| OT | Object Template |
| PS | Process-Structure |
| $PS_A$ | Abstract Process-Structure |
| $PS_E$ | Elementary Process-Structure |
| $PS_R$ | Reactive Process-Structure |
| TPO | Task Participating Object |

## 6. References

Asimov, I. (1991). Robot Visions, Roc (Reissue 5th March 1991), ISBN-10: 0451450647

Cao, T. & Sanderson, A. C. (1998). AND/OR net representation for robotic task sequence planning, In: *IEEE Transactions on Systems, Man, and Cybernetics - part C: Applications and Reviews*, 28(2)

Dario, P., Dillman, R., and Christensen, H. I. (2004). EURON research roadmaps. Key area 1 on 'Research coordination', Available from http://www.euron.org

Engelberger, J. F. (1989), *Robotics in Service*, MIT Press, Cambridge, MA, USA, 1st ed, 1989

Gostai. (2010). Urbi 2.0. Available from http://www.gostai.com

Gräfe, V. & Bischoff, R. (2003). Past, present and future of intelligent robots, *Proceedings of the 2003 IEEE International Symposium on Computional Intelligence*, In: *Robotics and Automation* (CIRA 2003), volume 2, ISBN 0-7803-7866-0, Kobe, Japan

IAT (2009). *AMaRob Project*, Institute of Automation, University of Bremen, Germany. Available from http://www.amarob.de

IAT (2011). *ReIntegraRob Project*, Institute of Automation, University of Bremen, Germany. Available from http://www.iat.uni-bremen.de/sixcms/detail.php?id=1268

Kampe, H. & Gräser, A. (2010). Integral modelling of objects for service robotic systems, Proceedings for the joint conference of ISR 2010 (41st International Symposium on Robotics) und ROBOTIK 2010 (6th German Conference on Robotics), 978-3-8007-3273-9, Munich, Germany

Kemp, C. C., Edsinger, A. & Torres-Jara, E. (2007). Challenges for robot manipulation in human environments, In: *IEEE Robotics and Automation Magazine*, vol. 14, pp. 20-29

Martens, C. (2003). Teilautonome Aufgabenbearbeitung bei Rehabilitations-robotern mit Manipulator - Konzeption und Realisierung eines software-technischen und algorithmischen Rahmenwerks, PhD dissertation, University of Bremen, Faculty of Physics / Electrical Engineering, (in German)

Martens, C., Prenzel, O. & Gräser, A. (2007). The rehabilitation robots FRIEND-I & II: Daily life independency through semi-autonomous task-execution, In: *Rehabilitation Robotics* (Sashi S Kommu, Ed.), pp. 137-162., I-Tech Education and Publishing, Vienna, Austria, Available from http://www.intechopen.com/books/show/title/rehabilitation_robotics

Microsoft. (2011). Microsoft Robotic Studio, Available from http://www.microsoft.com/robotics

Natarajan, S.K., Ristic-Durrant, D., Leu, A., Gräser, A. (2011). Robust stereo-vision based 3D-modeling of real-world objects for assistive robotic applications, in *Proc. of IEEE/RSJ International Conference on Robots and Systems (IROS), San Francisco, USA*

Ojdanic, D. (2009). Using cartesian space for manipulator motion planning - application in service robotics, PhD dissertation, University of Bremen, Faculty of Physics and Electrical Engineering

Prenzel, O. (2005). Semi-autonomous object anchoring for service-robots, *in B. Lohmann (Ed.), A. Gräser, Methods and Applications in Automation*, pp. 57 - 68, Shaker-Verlag, Aachen, 2005, ISBN 3-8322-4502-2

Prenzel, O., Boit, A. and Kampe H. (2008) Ergonomic programming of service robot behavior with function block networks, in *Methods and Applications in Automation*, Shaker-Verlag, pp. 31-42

Prenzel, O. (2009). *Process model for the development of semi-autonomous service robots*, PhD dissertation, University of Bremen, Faculty of Physics and Electrical Engineering

Quigley, M., Conley, K., Gerkey, B. P., Faust, J., Foote, T., Leibs, J., Wheeler, R., Ng, A. Y. (2009) ROS: an open-source Robot Operating System, In: *Proc. Of ICRA Workshop on Open Source Software*

Russel, S., and Norvig, P. (2003). *Articial Intelligence - A Modern Approach*, Prentice Hall, Upper Saddle River, New Jersey, 2nd ed.

Schlegel, C., and Woerz, R. (1999) The software framework SmartSoft for implementing sensorimotor systems, In: *Proc. of the IEEE/RSJ International Conference on Intelligent Robots and Systems (IROS)*, pp. 1610-1616

Schmidt, D. C. (2006). Model-driven-engineering, In: *guest editor's introduction*, pp. 25-31, IEEE Computer

Schreckenghost, D., Bonasso, R., Kortenkamp, D., Ryan D. (1998) Three tier architecture for controlling space life support systems, In: *Proc. of IEEE SIS'98*, Washington DC, USA

Simmons, R., Apfelbaum, D. (1998) A task description language for robot control, In: *Proc. of Conference on Intelligent Robotics and Systems*

Weld, D. S. (1999). Recent advances in AI planning, in *AI Magazine*, vol 20, pp. 93-123

Wright, R. S., Lipchak, B., Haemel, N. & Sellers, G. (2010). *OpenGL SuperBible: Comprehensive Tutorial and Reference* (5th Edition), Addison-Wesley, ISBN 978-0321712615

# 6

# Using Ontologies for Configuring Architectures of Industrial Robotics in Logistic Processes

Matthias Burwinkel and Bernd Scholz-Reiter
*BIBA Bremen, University of Bremen*
*Germany*

## 1. Introduction

The provision of goods and services accomplishes a transition to greater value-added-oriented logistics processes. The philosophy of logistics is changing to a cross-disciplinary function. Therefore it becomes a critical success factor for competitive companies (Göpfert, 2009). Thus logistics assumes the task of a modern management concept. It provides for the development, design, control and implementation of more effective and efficient flows of goods. Further, on aspects of information, money and financing flows are crucial for for the development of enterprise-wide and company-comprehensive success.

According to (Scheid, 2010a) this can be ensuring, by the automation of logistic processes. Based (Granlund, 2008), the necessity for automated logistics processes raises the focus on logistics by existing dominant factors of uncertainty and rapid changes in the business area environment. Therefore, the adoption of flexible automation systems is essential. Here robotics appears very promising due to its universal character as a handling machine. This is how (Suppa & Hofschulte, 2010) characterizes the development of industrial robotics: '[...] increasingly in the direction of flexible systems, which take over new fields with sensors and innovative fiscal and regulatory approaches.' Here, logistics represents a major application field. (Westkämper & Verl, 2009) describe the broad applications for logistics and demonstrate the capability for flexibility with examples from industry and research. Besides the technological feasibility, there is also the existing demand by logistics firms concerning the need for their application.

These representations demonstrate the interaction of robotics-logistics regarding the design of technical systems for the operator strongly driven by the manufacturer (technology push) and the technological standardization of the system. Robotic-logistics concentrates on the development and integration of products. Accordingly, standardization activities of the technical systems focus on components and sub-systems that represent the manufacturer-oriented perspective.

The main goal concentrates on the planning, implementation, and monitoring of enterprise-wide process chains of technological systems under the consideration of economic criteria. In this context, the interaction of the two domains 'process' and 'technology' are essential. Thus, the configuration design of technological layouts or machines is crucial. The harmonization of the two domains requires a systematic description framework concerning their exchange of information and knowledge. A high-level abstraction of knowledge representation in the description of the relationships and connections is essential. It also allows the description of implicit relationships such as comparative relationship notations. This applies to both qualitative and quantitative types of relationships. The outcome is a framework that is available to represent an object dependency between process and technology and to serve the described requirements for flexibility regarding logistics cargo, throughput and machine- and process-layout.

Thus, there is the need for qualitative description of relationship between process and technology by means of specific parameters and properties on a high-level abstraction.

## 2. Robotics-Logistics: Challenges and potentials

Since the 1970s, there has been a multifaceted development of the basic understanding of logistics. The origin of 'logistics' refers to the Greek 'logos' (reason, arithmetic) and the Romanesque-French ('providing', 'supporting'). In the past logistics were understood in delimited functions. Nowadays logistics are global networks, which are necessary to optimize. The understanding of the task itself changed from a pure functional perspective through process chains to value-adding networks:

Fig. 1 shows the historical development starting in the 1970s. Today's logistics is characterized by its value and integration in the appropriate process chains. The Federal Logistics Association designates logistic processes to the areas of procurement, production, distribution, disposal and transport logistics. (Arnold, 2006) designates differentiated performance-oriented processes as transport and storage processes. Storage processes are the processes of handling, order picking, and packing. Logistics services are evaluated based on delivery time, delivery reliability, inventory availability, delivery quality, and delivery flexibility. These are the objectives for both intra-logistics and extra-logistics. Logistics institutions, such as logistics service providers, provide value-added benefits to this process. These services are dependent of the collection and the output of their product 'commodity'. Finishing or outer packaging operations are examples here.

The logistics of the future will be essentially determined by the automation of material and information flows. In this area, automation systems in logistics already exist for several years. Application areas for these systems, such as de-palletizing and palletizing, sorting, and picking, are 'technically feasible and tested for decades' (Scheid; 2010b). The complete automation of the so-called intra-logistics is technically feasible. However, this situation is not encountered in practice due to the singular character of isolated applications. In future material flow technologies will be more modularized as (Straube & Rösch, 2008) identified. Modular automation systems maximize flexibility in the logistics systems and enable the re-utilization of technical components of handling and storage technology. To summarize the research requirements concerning these technologies (Straube & Rösch, 2008) ask for new modular constructions, which can combine different techniques based on their standardized modular features. This simplifies the integration into new systems. They describe a weakening tendency in new features for the components of conveying and storage technology. The focus is set on the configuration of system architectures composed of existing commercial components. This approach leads to process-specific integrated systems.

From an industrial point of view, multiple logistics areas display a high potential for the automation of processes (Scheid, 2010b). Thus, a high potential exists for the processes 'transport," 'storage' and 'de-palletizing." Transport processes will be automated in 2015 by nearly 30 percent. The reasons for the limiting borders for straightening the degree of automation are lying in the characteristics of the material and information flows. The existing process dynamics and process volatility are a handicap for standardized processes. The continuous automation of specific and individual processes appears to be difficult due to these reasons. Machine application requires great flexibility for adapting changing parameters.

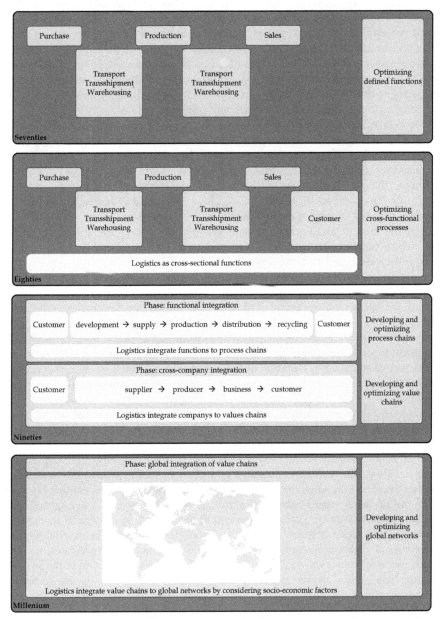

Fig. 1. Historical development of logistics philosophy [source: authors illustration following (Baumgarten, 2008)]

A fundamental role belongs to robotics. By definition, industrial robot systems are a central success factor in process automation due to their universality. The application of robotics systems in logistics factories should be designed flexibly. The automation of the processes under the customers' existing requirements can allow individually designed systems (Günthner & Hompel, 2009). In their recent study, the European Initiative (EUROP, 2009) identified the application of robotics in logistics as a central issue for the future. Thus, it highlights the broad range of application and diverse functions in this area. The operation of the systems under limited process standardization due to the complexity of the processes because of heterogeneous and manifold variables, leads to individual and special solutions in today's logistics factories. The adaption of the systems to changing process environment is hindered due to the process specific character of the systems.

(Fritsch & Wöltje, 2006) identifies the necessity for a paradigm shift from such individual system configuration and underline the relevance of standardized robot systems. This necessity establishes (Elger & Hausser, 2009) by describing the demand of more standardized solutions, which can also serve individual needs. The initiative (EUROP, 2009) characterizes the standardization of components and technical systems as an essential challenge for the so-called 'Robotics 2020." This concerns both hardware and software, and their interfaces among these components. In the authors' view, this requirement influences the system architecture essentially. In the view of the (EUROP, 2009), the system architecture accords robotics a central role. In the future architectures for robotic systems will be designed to both comprehensive configuration conditions and technical subsystems and components. They can be assigned from comparable and very different applications. Therefore, robotics systems will be more modularized in their architecture configurations in the medium term (until 2015). The interconnection between the modules is weakly configured in an overall perspective. On the one hand, this allows a rapid reconfiguration when changes of the process environment appear. On the other hand, the standardization of components and systems is besides the repeated partial usage the second driver of so-called 'adaptable configuration status." The long-term perspective for the year 2020 looks out for the development of architectures down to autonomous self-configurations.

The second crucial development is represented by the compositionality of robotic systems. A robotic system is compositional when the complexity of system architecture is based on compilation of the subsystems or components and their specific functions. The more sub systems or components will be used, the higher is the probability of complex system architecture. Thus, this configuration status is dependent on the process environment. This means that the robotic systems for self-changing or complicated processes must be explicitly designed to fit these requirements. The robotic system has to be configured process orientated. Robotics-logistics configuration conditions appear to diverge in comparison to the configuration condition of production robotics. This can be attributed to the characteristics of logistics processes. The process environment appears to have an essential influence on the technological configuration status of robotic systems. Out of the perspective of system theory, the degree of complexity can be influenced by its technological configuration status of the robotic systems and the characteristics of the process environment.

Thus, complicated processes often require robotic system architectures, which are composed of numerous components and are individually configured. This relationship can result in complicated or complex systems on the process- and on the technology-level. Additionally, procedural complexity influences technical complexity. The necessary reaction possibilities with technical components to procedurally dynamic events are the main driver here. The individual solutions counteract the intended economic standard solutions. Standardization serves to reduce complexity and have to integrate both the process environment and systems engineering. Robotic systems can be standardized by considering the two perspectives of the configuration.

The description of this relational structure is represented by an approach that works with a qualitative logical description on an abstract level. Current approaches to system modeling appear too formal. Ontological approaches with their level of abstraction are an interesting alternative. Despite the standardization of system architecture, a process-orientated configuration is to be ensured. The necessary flexibility intends to serve the dynamic and volatile processes. The construction and structure of the architecture has to be monitored and planned in its modular basic approach. To cover the historical, actual and future usage of technical systems, modular robotic systems are essential.

This book chapter describes a conceptually basic approach procedure for the representation of the relational structure between process and technology through an ontological vocabulary.

### 3. State of the art - modelling approaches for system representation

Examples of traditional modeling methods for representing systems where relationships between entities are described are: 'entity-relationship model,'' 'Petri nets,' and 'event-driven process chains' (Kastens et al., 2008, Seidlmeier, 2002, Siegert, 1996).

The Entity Relationship Model (ER-model) was developed in 1976 by Chen. It allows delimited systems to be represented in a way which is intelligible for all involved. The entities (objects) and the relationships between the objects form the basis of the modeling. Regarding the purpose of the modeling, only objects, relationships, and attributes are described (Chen, 1976). The method of Petri nets represents structural coherence between sets of events (Kiencke, 1997). In general, a Petri net is a graphic description where the transaction of the generation of sequences of event-driven networks is represented. It consists of types of nodes, which are representative of a so-called position or transition conditions and events. A directed edge connects a position with a transition. Petri nets are capable of describing a large class of possible processes (Tabeling, 2006). Event-Driven Process Chains (EPC) modeling is a process-oriented perspective on functions, events, organizational units, and information object systems. A process chain is defined by modeling rules and operators (Staud, 2006).

Systems can be also modeled by using ontologies. The concept of ontology originates from philosophy and describes the 'science of being." Many authors define ontology from different perspectives. (Gruber, 1993) describes ontologies as the explicit specification of a conceptualization. The abstract level has the advantage that many basic approaches of different research areas are defined. For example, linguistically and mathematically oriented ontologies are combined due to this definition. (Stuckenschmidt, 2009) establishes the common reference to this definition by many authors. (Studer et al., 1998) take it as a basis and defines ontologies from their formal logic: 'An ontology is a formal, explicit specification of a shared conceptualization.' They emphasize the machine-readable formality of ontology. (Neches et al., 1991) specifies this idea and describes ontologies as 'basic terms and relations comprising the vocabulary of a topic area, as well as the rules for combining terms and relations, to define extensions to the vocabulary.' According to this understanding, concepts are defined through basic distinctions of objects and their rule-based relationships to each other. (Bunge, 1977) describes ontology as the only area of science besides the fields of natural and social sciences, which focuses on concrete objects and concrete reality. Ontologies are to be assigned based on philosophy since they stress the basic principles of virtual science, which cannot be proven or refused by experiments.

Ontologies represent knowledge, which is structured and provided with information technologies. They can be a crucial part of knowledge management. According to (Staab, 2002) knowledge management has the goal to optimize the requirements for employees' performance. The following factors 'persons", 'culture", 'organization" and 'basic organization processes" are the major success criteria for knowledge management. According to (Gruber, 1993) ontologies can facilitate the sharing and exchange of knowledge.

There are many kinds and types of ontologies. Depending on their internal structure, ontologies vary in their complexity, as represented in Fig. 2:

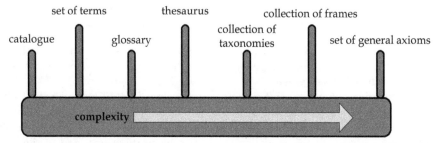

Fig. 2. Types of ontologies organized by increasing complexity [source: authors illustration following (Herb, 2006)]

Examples for trivial complex ontologies are simple catalogues or collections of concepts. Maximally complex ontologies contain an amount of general and weak-defined axioms. An interesting type is taxonomies, which can be defined as a hierarchical classification of concepts in categories.

Taxonomies are also considered as an attenuated definition of ontology. According to (Herb, 2006), they include a series of concepts that are interlinked by hereditary structures. Depending on their nature, ontologies can be applied and re-applied with different levels of intensity (Gómez-Pérez, Fernández-López, Corcho 2004). Ontologies can be classified in so-called 'lightweight ontologies' and in 'heavyweight ontologies.' 'Lightweight ontologies' describe notions (concepts), taxonomies, and relationships and properties between terms. Additionally to these properties, 'heavyweight ontologies' also consider axioms and constraints.

The ontological modeling of systems is possible through the application of existing relationships and rules. (Steinmann & Nejdl, 1999) describe the two tasks of ontologies. In the first sense, ontologies describe the nature of the constituents and the principles. He designates these as 'grammar of reality'. In the second sense, ontologies establish the objects and connections, which (Steinmann & Nejdl, 1999) designate as the 'encyclopedia of reality'. In the first sense, they function as meta-models. Abstract modeling concepts are described and provide the framework for the ontology. Specific meta-model-oriented ontologies will be designated as representation ontologies. For example, the frame-ontology in Ontolingua can be mentioned here, according to (Gruber, 1993). The ontology provides a grammar composed of concepts, relations and attributes. In the second sense, ontologies describe conceptual models are based on structures and correlation of the area of a specific application. Examples of existing conceptual models are legal texts, integration of application systems or open systems.'

Comparing classic and ontological methods, some differences can be identified. Ontologies describe the composition of reality. Traditional modeling approaches assume this information to be known. In this context (Herb, 2006) ascertains, that ontologies are applied for concept-based structuring of information. In his view, ontologies are essentially for more detailed structured information than conventional sources. (Stuckenschmidt, 2009) describes the existence of objects and items and the representation form. (Steinmann & Nejdl, 1999) detail this approach and describe it as a central factor for understanding of items. He concludes that ontologies always are based on a highly abstract level in comparison to model-based approaches.

The authors also indicate the borders of ontological modeling. The crucial difficulties are inconsistencies in classification of meta-data in ontologies, application of meta-data and the distinct classification and structuring of information. Therefore, these aspects are attributable to the highly abstract level of ontologies. Abstract notations lead to such assignment, classification, and structuring issues.

Ontologies can be differentiated in two aspects, conceptual and formal logical nature. According to (Swartout & Tate, 1999), the first aspect has the task of depicting and composing structures of reality. The second addresses the creation of the semantic framework with the definition of objects, classes and contexts. There are many basic approaches to different ontologies in the literature, oriented to the areas of application. (Bateman, 1993) describes the existence of basic types of interconnected entities and describes the so-called 'design patterns.' These are entities that can be differentiated according to the types 'endurant', 'perdurant/occurrence,' 'quality' and 'abstract.' While entities of the 'endurant' type have a continuous and predictable nature, entities of the type 'perdurant/occurrence' describe events that occur unexpectedly and unpredictably. Entities of the types 'quality' and 'abstract' unite properties, attributes, relations, and comparatives.

The ontology DOLCE is an example of the application of these basic types. DOLCE was developed by the Institute of Cognitive Science and Technology in Trento, Italy and stands for 'Descriptive Ontology for Linguistic and Cognitive Engineering.' DOLCE attempts to impart meanings to things and events. Here, entities deal with the meanings through use of agents in order to obtain consensus among all entities regarding to the meaning. (Gangemi et al., 2002) treated this principle in a plausible way. Further examples of conceptual ontologies are WordNet, the 'Unified Medical Language' ontology, 'Suggested Upper Merged Ontology," the ontology of 'ε-Connection' (Kutz et al., 2004), the ontology of 'Process Specification Language,' and the ontology 'OntoClean."

Another key component of conceptual ontologies is ontology engineering. Ontology engineering is concerned with the process design of ontology development, in order to create and to apply ontologies. There are multiple methods here. (Wiedemann, 2008) lists these as follows:

- Ontology Development
- Ontology Re-Engineering
- Ontology Learning
- Ontology Alignment/Merging
- Collaborative Ontology Construction

Ontology Development deals with the question of methodological development and the composition of ontologies. Ontology Re-Engineering focuses on existing approaches and adapts them to the current task. Ontology Learning focuses on approaches for semi- or fully-automatic knowledge acquisition. The Collaborative Ontology Construction issued guidelines for the generation of consensual knowledge. Ontology Merging combines two or more ontologies in order to depict various domains. This method allows handling knowledge that is brought together from different worlds.

(Gruninger, 2002) describes formal logical ontologies as communication, automatic conclusion and representation and re-utilization of knowledge. Formal logical ontologies aim to depict a semantic domain through syntax. The concept of semantics is to be classed in semiotics and describes the theory of signs. Semantics can be also defined as the 'theory of the relationships among the signs and the things in the world, which they denote' (Erdmann, 2001). Semantics are relevant for formal logical ontologies for modeling and generating calculations on a mathematical foundation. This basic approach with its syntax performs a key relevance by providing the mathematical grammar and the concretely denotable model. Exemplary syntaxes are algebraic terms, logical formulas or informational programs. Formal logic provides a language for formalizing the description of the real world and the tool for representing ontologies. It is differentiated according to propositional logic and predicate logic. In propositional logic, there exist exactly two possible truth-values: true or false. Predicate logic consists of terms and describes real world objects in an abstract manner by means of variables and functions. (Stuckenschmidt, 2009) presents methods and techniques of the notation.

Formal logic ontologies do not allow automatic proofs. Only computer-based evidence for sub-problems is possible. Examples of formal logical ontologies are OntoSpace, DiaSpace, OASIS-IP, CASL, OIL, and OWL.

In summary, it can be stated that both ontology types can be classified in different types according to (Guarino, 1998): 'top-level ontologies," 'domain ontologies' and 'application ontologies" which already represent known data and class models. 'Top-level ontologies' describe fundamental and generally applicable basic approaches which are independent of a specific real world. Their level of abstraction is high that allows a wide range of users.

'Domain ontologies' focus on a specific application area and describe these fundamental events and activities by specifying the syntax of 'top-level' ontologies. 'Application ontologies' make use of known data or class models which apply to a specific application area.

The following table finally summarizes the described ontologies and compares them according to the presented properties and characteristics. Furthermore, the relevance of ontologies applicable for robotic logistics is specified and the ontologies of 'Process Specification Language' and the ontology 'OntoClean' are highlighted:

| ontology | author | application area | characteristics | relevance |
|---|---|---|---|---|
| DOLCE | Institute of cognitive Science a Technology, Italien | semantic Web | cognitive basis | partially relevant; wide knowledge, uses cognitive aspects |
| Onto Clean | Laboratory for Applied Ontology, Trento | hierarchical strucutre of knowledge | checking inconsistencies automatically | relevant for structuring processes and robotics technologies |
| PSL | National Institute of Standards and Technology, Gaithersburg | neutral representation of process knowledge | modular d | relevant for describving relations between processes and robotics |
| WordNet | Princeton University | representation of natural languages in IT applications | lexical database | not relevant |
| UMLS | National Libary of Health | database for communicating medical terminology | terminology for medical applications | not relevant; especcially for biomedicine |
| SUMO | Teknowledge Corporation | providing information in databases and in the internet | combined out of multiple ontologies | not relevant; designed for automated verification |
| E-Connection | University of Liverpool | complex correlation of different domains | connecting different domains in a formal and logical way | relevant; complex |
| CASL | Common Framework Initiative | first-Order-logics for subsuming specific languages | modular concept | not relevant; formal approach |
| F-Logic | Stony Brook University | deductive database | conceptional and object oriented | not relevant, formal approach |
| OIL | Vrje Universität, Amsterdam | web based language | formal infrastructure for semantic web | not relevant, formal approach |
| Ontology Web Language | World wide Web Consortium | representation of correlation in the semantic web | base for integration of software | not relevant, formal approach |

Table 1. Comparison of selected ontologies in the context of Robotic Logistics [source: author's illustration]

## 4. Logical ontologies for configuration of individual system architectures

### 4.1 Required ontological framework

'Robotic-Logistics' formulates the central expectations to the ontology for configuration robotic system architectures. The input and output variables of the environment due to the reference process have to be defined. On this basis, the relevant domains 'technology' and 'process' can be described as to contents. Classes and variables structure them. On the process side, the reference process is addressed. In this domain, the direct upstream and downstream processes of the reference process are also relevant. The output of the upstream process provides the input of the reference process. The output of the reference process provides the input of the downstream process. The relevant technical systems and components of robotic-logistics will be structured in the technology perspective. The regulatory framework has described the following entities. Fig. 3 gives an overview of the hierarchical structure of the domains 'process' and 'technology':

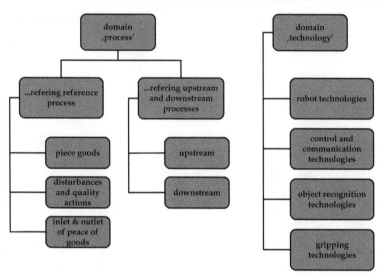

Fig. 3. Class structure of the 'process' and 'technology' domains [source: authors illustration]

The entities in these two class structures are the main processes of meta-model, which will be presented in this chapter. On this basis, process modules and process elements of the reference process are derived with application of the regulatory framework.

The reference process is situated in its systemic environment. It influences the reference process with input and output variables. Thereby the input describes the general framework and restrictions, which are valid for the robotic system. The output results directly from the target dimension of the automation task. The parameters cover technical, organizational, and economic aspects. (Kahraman et al., 2007) define a multi-criteria system for the evaluation of robotic systems, which provides a multiple key factors for the evaluation of robotic systems.

With the development of entity structures of the two domains of the reference process and the inputs and outputs of the environment, the fundamentals of ontology development are set. Based on these structures the hierarchical ontological taxonomies are created with the aid of the ontology OntoClean. This is necessary in order to be able to describe the relations between the two domains through ontology, the Process Specification Language.

## 4.2 Conceptional ontology for descriptive process technology relations

This section introduces a two-stage approach. In the first phase, the hierarchical structures of the respective domains are composed. The procedure model of (Stuckenschmidt, 2009) offers advantages for the creation of these taxonomies. This approach forms the taxonomies through the OntoClean ontology and analyzes potential sources of error. In the lowest level of taxonomy elements, properties and attributes of the process elements are denoted. They define the reference process. Thus, for example, the process module 'piece goods' with the process element 'bulk" displays the property 'five kilos". Due to this definition, the reference process is individualized and specified.

The second phase provides the combination of the two domains. The description of these relations is done through the ontology of 'Process Specification Language.' It is based on the descriptive notation of functions and processes through its manifold concepts and relations in different levels of detail. Each participant in a pair of relationship is standardized and the relationship is jointly depicted. Due to the functional and procedural point of view of the ontology, the relationship can be well illustrated. The representation is done by focusing on process elements of one domain that cause an impact on the process elements of the second domain.

For preparation, the conceptual framework is defined as the delimitation of the considered environment to be covered. It is defined according to the procedure model developed by (Figgener & Hompel, 2007). They describe a regulatory framework for reference processes:

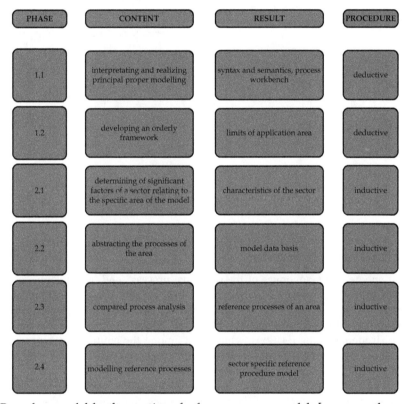

| PHASE | CONTENT | RESULT | PROCEDURE |
|---|---|---|---|
| 1.1 | interpretating and realizing principal proper modelling | syntax and semantics, process workbench | deductive |
| 1.2 | developing an orderly framework | limits of application area | deductive |
| 2.1 | determining of significant factors of a sector relating to the specific area of the model | characteristics of the sector | inductive |
| 2.2 | abstracting the processes of the area | model data basis | inductive |
| 2.3 | compared process analysis | reference processes of an area | inductive |
| 2.4 | modelling reference processes | sector specific reference procedure model | inductive |

Fig. 4. Procedure model for the creation of reference process models [source: authors illustration following (Figgener & Hompel, 2007)]

The aspects of an application area are defined. Thus, displayed in fig. 4, six phases for the generation of a reference model are described. For the existing problem phase 1.2, phase 2.1, and phase 2.2 are especially relevant. Phase 1.2 describes the regulatory framework. The process modules and process elements are defined in phase 2.1 and 2.2. This distinction allows the reduction and control of the complexity and expenditure for model creation through the ontology.

With these results, both phases of the ontology model can be completed.

Fig. 5 shows the interdependence of both ontologies:

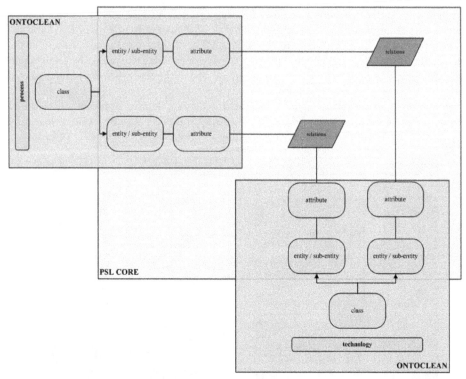

Fig. 5. Spheres of action of the ontologies 'OntoClean' and 'Process Specification Language'[source: author's illustration]

In the first sphere, the taxonomies of the domains 'process" and 'technology" are defined with 'OntoClean'. Depending on the reference process, the process taxonomies will be characterized by properties. They customize the taxonomies. The fig. also displays the sphere of action of the first part namely 'PSL Core' of the ontology 'Process Specification Language." It identifies the relations, which exist among the entities and properties of the process domain and the entities and attributes of the technology domains. Summarized the figure works out 2 phases of the ontology model.

The first phase develops the taxonomies using OntoClean. The taxonomies are denoted and structured. Based on the notation of the meta-properties, the accuracy of the taxonomies is analyzed. Inconsistencies regarding the clearness of the hierarchies arise when relations are utilized incorrectly. This leads to incorrect and misleading interpretations of the ontology (Herb, 2006). The OntoClean process examines existing subsuming structures existing between classes by using meta-features.

The second phase depicts relations between the domains by using the ontology of 'Process Specification Language'. Fixed taxonomies for the domain 'process" and the domain 'technology" for a specific reference process, are the basis for representation of the interaction between the two domains. The main goal of the ontology is to figure out parameters or components of a domain affecting the second domain. Besides the demonstration of the existing relations, the description of the quality of the relationship is an essential aspect. Thereby both the direction of the relationship and the qualitative description will be identified. With the ontology model, the defined requirements will be satisfied as follows:

| requirement | ontology | contribution |
|---|---|---|
| structuring domains | OntoClean | - identification of relevant terms<br>- structuring by using taxonomies |
| notation relations | PSL | - identification and notation of relations between process and technology |
| description relationship | PSL | - description of the relations |

Table 2. Handling the requirements through the ontology model [source: author's illustration]

The ontology OntoClean structures the domains. It defines the concepts and composes the taxonomies of the domains 'process" and 'technology". The 'Process Specification Language' note the relations between the parameters. The worked out ontology model is the basis for the individual process modularization and configuration of technical robotic systems.

### 4.2.1 Definition of taxonomies using the ontology 'OntoClean'

The structuring of the domains is done by defining taxonomies. The usage of taxonomies joins and collects concepts and entities and forms a base frame for these ontologies by structuring them. Here, the relationships are developed associatively mutually. Descending rules work out the taxonomy structures. Due to the qualitative character, the taxonomies are often incorrectly distinguished. The process of 'OntoClean' creates taxonomies and checks the consistency and accuracy of the structures.

The procedure involves the definition of taxonomies and their meta-properties. It aims at overcoming the frequent deficit of false descent of entities in the taxonomic structure. These erroneous subsuming structures will be avoided by a philosophy-based distinction of the entities and classes with meta-properties. (Herb, 2006) describes comprehensively the meta-properties 'identity,' 'essence and rigidity,' 'dependency,' and 'unity.' Using these meta-properties, the taxonomies are distinctly defined through the concepts of class, entity, instance, and property. Entities describe the objects of taxonomy, which are collected in classes. Entities, which have a common property, instantiate a class and will be defined as instances. This is of great relevance. Especially the representation is challenging due to multiple components and parameters, which are displayed in both the domain 'technology' and the domain 'process'. Table 3 provides an overview of the conceptions.

The concepts are the basis of the taxonomies to be created. They are based on the entity structures. For a specific reference process, the entities are reviewed and adapted individually. The procedural taxonomies are developed based on the meta-models of process standardization developed by (Figgener & Hompel, 2007). Depending on the type of process, the main processes, process modules and process elements are applied. Based on the structured system techniques in the domain 'technology', the technical taxonomies are defined due to the commercial state of the art.

The claim of universality is not maintained. The conception is defined to each reference process specifically. This increases the risk of erroneous and inconsistent definition and description of the concepts. Due to this circumstance, the analysis and validation of the developed taxonomies is an essential part of ontology development.

| term | definition of procedure model due to fig. 4 | commentary |
|------|---------------------------------------------|------------|
| class | main process | structuring of classes |
| entity | process module | entities which are subsumed in one class |
| instance | process module with same attributes | all entities with same attributes in one process module |
| property | process element | characterization and individualization of the reference process |

Table 3. Definition of conceptions in the framework of 'OntoClean' [source: authors illustration]

The OntoClean process provides a procedure of subsuming, shown in fig. 6:

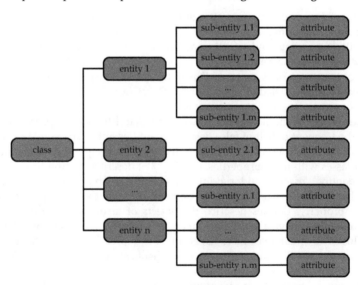

Fig. 6. General construction of taxonomy [source: author's illustration]

A class A subsumes a class B if all instances of class B are always also instances of class A. Fig. 6 presents a class with n entities. Each entity can display further lower-level hierarchical entities, known as sub-entities. Thus, the number of vertical levels is unlimited. On the lowest vertical level, the reference process is individualized by distinct properties. They provide the specific information about the reference process. These may be quantitative or qualitative. As an example here, for a procedural taxonomy, a sub-entity of type 'mass' can be specified with the quantitative property of '22 kg'.

A special case is presented due to the class structure 'environment'. Here, both are structured the environmental framework conditions and the target dimensions. This taxonomy is independent of the reference process, and provides an example, shown in its basic structure, as follows:

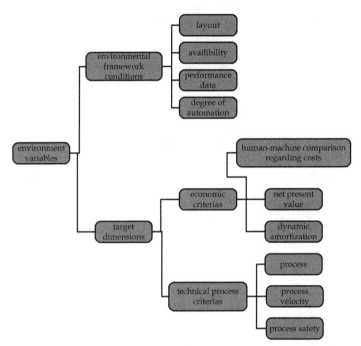

Fig. 7. Taxonomy of the environment variables [source: author's illustration]

The environmentally framework conditions define the technical requirements, such as availability and performance data. The listed properties are specifically defined for each reference process. They customize the process. By means of the target dimensions, technical and economic criteria are carried out. In this context, process safety or process velocity on the technical side are determined. On the economic side, capital value or amortization time is identified. The target dimensions represent the criteria for success of the realization of the robotic system in an ex-post manner.

For each taxonomy and its class K, a notation is defined with the property M. K is denoted as +M, when M applies to all instances of K. The notation -M is used, if not all instances of K have the property M. If M is not valid for any instance of the class K, this relationship is denoted by ~M. Each of the four meta-properties will be reviewed to that effect for each class and entity. This describes the meta-property, 'essence and rigidity' by fixing essence in general first. Second, the specific form of the essence, the rigidity, is described. A property is essential for an entity, if it occurs in every possible situation in the entity. In the next step, a property is rigid, if it is essential for all instances (+R). Non-rigid properties are referred with –R. They describe properties for those entities, which are needed not, but may be instances of the class. Anti-rigidity (~R) is available if there is no instance of associated class instance of the corresponding class.

The meta-property 'identity' describes criteria, which distinctly identify classes and differentiates instances from each other. Both classes and upper classes can provide these identity criteria. The upper classes inherit the criteria. In the first case, the classes are marked with +I. Thus, the identity criterion has been inherited by an upper class. In the second case, the criterion of identity is first defined in an upper class and is marked with +O. Classes that require a further identity criterion as restrictions for distinct definition are denoted with –I.

The third meta-property 'unity' is related to the property 'identity' and describes the affiliation of certain entities to a class. A unity criterion defines a unifying relation of all entities, which are interconnected. The corresponding classes are distinguished with +U. ~U denotes those entities of a class that cannot be distinctly described. If there is no unity criterion provided, the class is described with –U.

The fourth meta-property 'dependence' describes the dependence of a class to other. This fact is relevant if an instance of a class may not be an instance of a second class. Dependent classes will be listed with +D, while independent classes are notated with –D.

In summary, the meta-properties are defined as follows, according to (Herb, 2006):

| Meta-property notification | definition |
|---|---|
| +R | a property is essential for all valid instances |
| -R | a property that has not inevitable an entity that is an instance of its class |
| ~R | a property where an instance of an allowing class belongs to an instance of a regarded class |
| +I | classes which differentiate due to the criteria of the allowing instances |
| -I | class that does not have an identity criteria |
| +O | identity criteria that is defined for the first time and is not transmitted |
| +U | unity criteria that denotes connected entities |
| -U | none unity criteria is existing |
| ~U | connection of entities which cannot described definitely |
| +D | dependent classes |
| -D | independent classes |

Table 4. Summary definition of meta-properties, [source: authors illustration following (Herb, 2006)]

The review of meta-properties shows incorrect taxonomy structures and makes their correction possible. The next step involves reviewing the consistency of the meta-properties with each other. This will determine whether there are inadmissible combinations of meta-properties. , An example for such a combination is +O und –U. The next step focuses at the removal of all non-rigid classes from the taxonomy.

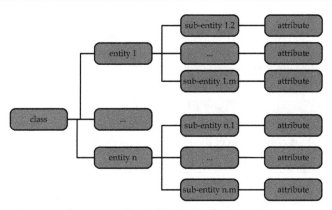

Fig. 8. Backbone taxonomy [source: author's illustration]

That figure points out for an exemplary illustration the removal of the non-rigid 'sub-entity 1.1' and the non-rigid 'entity 2'. This procedure results in the so-called backbone taxonomy. In the next step, subsuming structure has to be examined. It checks any violations of subsuming restrictions. Subsuming is described with the relation 'is-a' and is visualized by arrows. For instance, further relations are described with the notation type 'has.' To avoid false distinctions the arrows are inscribed with the relation name. The hierarchies can be described in the following ways:

- *Have*: The relationship type 'have' connects an attribute with a concept. Thereby, the Attribute is a type of the concept.
- *Att*: This type of relationship describes properties of elements. A concept can take on several properties simultaneously. They do not need to be met simultaneously at all.
- *Is a*: The relationship describes traditional subset relations (subsuming relations).

In the last step, the non-rigid classes and entities are added. Within this last step, the taxonomy is completed.

### 4.2.2 Description of the interaction by using the ontology of 'Process Specification Language'

In this section, the interacting entities and attributes have to be identified between the two domains by using the created taxonomies. In order to give these interactions a qualitative meaning, (Schlenoff et al., 1999) propose an approach to denote interactions of processes in independents worlds. Therefore, he develops the terminology 'Process Specification Language' (PSL). PSL is a neutral, standard language for process specification for the integration of multiple process applications within the product life cycle. The language is versatile in application and uses multiple concepts, classes, and functions to describe complex processes. Through its manifold applications and many years of further development, the language has diversified and expanded. PSL consists of several modules. The principle fundamental concepts are set in the first module which is called 'PSL-Core'. The module provides four concepts with corresponding functions. According to (Schlenoff et al., 1999), the aim of this module is to fix axioms to describe trivial process connections using a set of semantic relations.

The description of further and more complex processes is carried out with other modules, the so-called extensions. PSL offers in total three extensions: 'outer core', 'generic activities' and 'schedules'. The module 'PSL outer core' deals with generic and broadly based concepts regarding to their applicability. The module 'generic activities' defines a terminology to describe generic activities and their relations. The module 'schedules' describes the application and allocation of resources to activities under the premise of satisfying the temporary restrictions:

| term | definition |
|---|---|
| PSL | short notification of the ontology 'Process Specification Language' |
| PSL module | group of concepts of the PSL |
| relation | interrelation between an entity couple of 'process domain' and 'technology domain'. None of the entities is a ad-hoc activity. |
| activity | process or technical entity or sub-activity that is continuous and relates to the second domain. |
| ad-hoc activity | process or technical (sub-) entity including its attribute that existence is not calculable |
| concept | first and highest level of a PSL |
| class | second level of PSL |
| function | third and lowest level of PSL |
| ad-hoc relation | relation of an entity couple of the process and technology domain. Minimum one activity is an ad-hoc activity. |

Table 5. Definition of relevant terms of PSL [source: author's illustration]

The definitions are based on the adaptation of the ontology to the current requirements. With these concepts, the individual concepts of this approach will be presented and adapted to this task. With the creation of taxonomies, the relations of taxonomy properties of both domains are identified and described.

This section describes the identification of existing relations and their corresponding notation using the vocabulary of the first module 'PSL Core' of the ontology 'Process Specification Language'. The module contains three concepts. The first concept 'activity' describes general activities, which appear to be predictable and manageable. They do not have to be determined detailed. For instance, 'activities' may be standard processes of a recurring nature. The concept exhibits two different types of functions for further concepts. Function one focuses on further planned activities ('activity') ('is-occurring-at'). Function two describes the connection to unpredictable and unplanned activities ('occurrence-of').

The second concept, 'activity occurrence' describes a unique activity that proceeds unforeseen and unplanned. The concept can also exhibit two different functions for other concepts. The function 'occurrence of' is analogous to the second function of the first concept and describes the initiation of a second type of unpredictable activity of the type 'activity occurrence.' The second function describes the relationship to a concept of the type 'object'. This function expresses the impart of the concept 'activity occurrence' with a none further defined significance to the second concept named 'object.'

The third concept, 'object' describes all activities which do not correspond to any of the above concepts. The concept has two functions. The first relation of the type 'participate in' describes the concept 'object' which receives a non further defined relevance for a concept 'activity.' The second relation 'exists-at' describes an existing relevance to a particular point of time.

The entities are, inclusive of their properties, distinguished from the taxonomies of process and technology domains with these concepts. Here, procedural entities and properties can exist which are either calculable or definable. These activities relate to the concept 'activity.' Unpredictable, indefinable or changing conditions can be described as ad-hoc activities and assigned the concept of 'activity occurrence.' Other logistical or technical objects are called objects and assigned to the concept 'object.' An example describes the entity 'general cargo' as an activity (code 1.1) with its property 'cubic' and the entity 'stock situation' for a concept named ad-hoc activities (code 1.2) with the property 'chaotic'. An example of an object (code 1.3) is a technical process such as the process of recognizing the cargo. The following table summarizes the results of the relevant vocabulary:

| PSL module | concept | definition | relation | definition | modification for robotics-logistics | code |
|---|---|---|---|---|---|---|
| PSL Core | | | | | | 1 |
| | acitvity | a general non-defined acititvity | | | defined and calculable activity that notes an entity or an attribute of taxonomy. | 1.1 |
| | | | is-occurring-at | a primary activity generates a secondary activity at a defined time | the concept 1.1 generates a concept 1.1 | 1.1.1 |
| | | | occurrence-of | the primary concept generates a secondary non-expected activity | the concept 1.1 generates a concept 1.2 | 1.1.2 |
| | activity occurrence | a temporary activity and specific activity that occurs nonrecurring | | | a non-calculable and changing ad-hoc-activity that notes an entity or attribute of a taxonomy | 1.2 |
| | | | occurrence-of | the primary activity generates a secondary non-expected activity | the concept 1.2 initiates a new concept 1.2 | 1.2.1 |
| | | | participates-in | a primary activity generates a non-definable relevance for an object | the concept 1.2 generates a non-defined relevance for an object at a specific time | 1.2.2 |
| | object | all entities that are not an activity or activity occurrence | | | entity or attribute of a taxonomy that are not concepts 1.1 or 1.2 | 1.3 |
| | | | participates-in | a primary activity assigns a non-defined relevance to an object in a specific time | the concept 1.1 assigns a relevance to a concept 1.3 | 1.3.1 |
| | | | exists-at | an object exists to a specific time | the concept is relevant in a specific time | 1.3.2 |

Table 6. Vocabulary PSL module 'Core' [source: author's illustration]

In a first step, the implementation of ontologies for a specific reference process is associated with concepts and properties of the valid entities. The second step identifies and denotes the relations between the concepts. A matrix representation is provided which is shown in the general structure in tab. 7. The columns show the entities and properties of the technology domains. The lines depict the process domains. The individual hierarchy steps of taxonomies are presented. As described, they were indicated by the hierarchic structure. The coding of the lines and columns indicates the respective levels of the hierarchic structure. Additionally, the identified concepts of the respective sub-entities and properties are noted on the lowest structural level. In the cells, the interaction from tab. 6 are noted and distinguished by means of the coding. For example, the procedural sub-entity 1.1.1 affects the technical components 1.1.1 through the relationship 'object-participates-in' (code 1.3.1).

| | | | | Code | T.1 | T.1.1 | T.1.1.1 | ... | ... | T.n | T.n.m | T.n.m.o |
|---|---|---|---|---|---|---|---|---|---|---|---|---|
| Process Specification Language | | | | technical- taxonomy | system technique 1 | | | | ... | ... | system technique n | |
| | | | | | | system technique 1.1 | | | ... | ... | | system technique n.m |
| PSL Core | | | | | | | component 1.1.1 | ... | ... | | | component n.m.o |
| Code | process taxonomy | | | PSL-concept | | | concept 1.z | ... | ... | | | concept 1.z |
| P.1 | Class 1 | | | | | | | | ... | ... | | |
| P.1.1 | | entity 1.1 | | | | | | | ... | ... | | |
| P.1.1.1 | | | sub-entity 1.1.1 | concept 1.x | | | 1.3.1 | ... | ... | | | 1.x.y |
| ... | ... | ... | ... | ... | ... | ... | ... | ... | ... | ... | ... | ... |
| ... | ... | ... | ... | ... | ... | ... | ... | ... | ... | ... | ... | ... |
| P.n | class n | | | | | | | | ... | ... | | |
| Pn.m | | entity n.m | | | | | | | ... | ... | | |
| P.n.m.o | | | sub-entity n.m.o | concept 1.x | | | 1.x.y | ... | ... | | | 1.x.y |

Table 7. General matrix representation of the process-technology relations in accordance with PSL module 'core' [source: author's illustration]

The vocabulary allows the description of the relational structure for a dedicated reference process, which describes the relations among the procedural entities and the technical components.

### 4.3 Industrial application: Depalletizing plastic boxes with a robotic system
This section presents the robot based automation of a simple industrial application by using the presented ontological framework. The presented example focuses on the interaction between 'piece good' of the 'process domain' and 'gripper' of the 'technology domain'. Here the automation of a logistics process by using the ontological framework will be presented. Online books shops package their goods in plastic boxes. Logistics Providers handle these boxes for delivering to the customer. Hence, the boxes are send on pallets in swap bodies by using trucks. The logistics provider unloads the trucks and imports them in their distribution center which operates with a high degree of automation. Therefore the boxes have to be depalletized and brought onto the conveyor technology system. In general this separation is done manually. A robotic systems was configured and integrated by using the ontological framework to automate this reference process. The following figure displays the process with the implemented robotic system:

Fig. 9. Industrial application of a robotic system for depalletizing plastic boxes: result of the ontological configuration [source: author's illustration]

The illustration points out that the configuration of the technical system depends on the parameters of the process. The upstream conveyor supply box pallets including a buffer function. The downstream conveyor conveyors the single boxes into the distribution cycle. The task of an robot-based automation system focuses on the handling of single or multiple boxes and the lay down onto the roller conveyor. Using the presented framework for configuring the robotics architecture, the first step of generating the procedural and technical taxonomies has to be executed. The class structure with its entities and attributes of the 'process domain' can be assigned with attributed as followed:

| Entity | meta- property | sub-entity | attribute | meta-property |
|---|---|---|---|---|
| geometry | +R, +I, -U, +D | form | cubic | +R, -I, -U, +D |
| | | dimension (min, max) | length = 55 [cm] width = 40[cm] height = 30 [cm] | +R, +I, -U, +D |
| | | volume | V = 27 [l] | -R, -I, -U, -D |
| | | surface | closed | +R, -I, -U, -D |
| material | +R, -I, -U, -D | art | plastic | +R, +I, +U, +D |
| | | stability | high | -R, -I, -U, -D |
| packaging | +R, +I, -U, +D | strapping | 1 | -R, +I, -U, -D |
| | | type of packaging | single | -R, +I, +U, +D |
| mass | +R, +I, -U, +D | weight | 28 [kg] | +R, +I, -U, +D |

Table 8. Entities and attributes of the class structure 'piece goods' of the 'process domain' [source: author's illustration]

The table displays the various entities with its attribute regarding the class structure 'piece goods'. For instance, the meta-properties define the taxonomy of this class. Also the hierarchical structure is defined. For instance, incorrect assignments of sub-entities will be avoided. Additionally the table assigns relevant attributes of the reference process to entities. The meta-properties figure out that the sub-entities 'geometry' and 'mass' are quite important due to their essential (+R). Furthermore they give identity to their class (+I). Finally, the corresponding taxonomy is presented in figure 10:

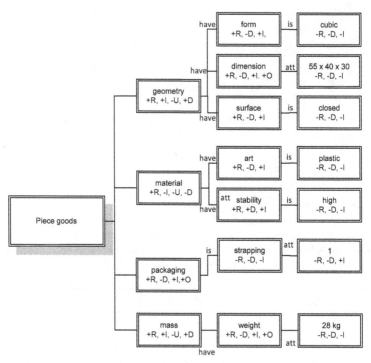

Fig. 10. Backbone taxonomy of the class structure "piece goods" of the "Process domain" [source: author's illustration]

It displays the corrected taxonomy of the exemplary class structure of the industrial application. The class "piece goods" consists of four entities (2nd level) and seven sub-entities. These are related with to attributes. The relations 'subsumption' (is) or 'attribute' (att) describes the connection to the entities. Subsumption are given if the attribute is part of the sub-entity.

The following step focuses on interactions between the "process domain" and "technology domain". Therefore, the relations are noted. Table 9 exhibits the relationship to the system technology "robotics" which unite the entities "kinematics", "geometry", "load", "accuracy" and "installation". Table 9 displays the relation codes between the two domains. For instance, there are some relations of the type "ad-hoc-activity" and "activity". For instance, the strapping has an influence to the accuracy of the robot. Also the mass defines the type of the robot. The table resumes these relations:

| Process Specification Language | | T.1 robotics | T.1.1 | T.1.2 | T.1.3 | T.1.4 | T.1.5 |
|---|---|---|---|---|---|---|---|
| | PSL Core | | kinematics | geometry | load | accuracy | installation |
| P.1 | piece goods | | | | | | |
| P.1.1 | geometry | | 1.1.2 | 1.1.1 | | | |
| P.1.2 | material | | 1.1.1 | | 1.2.1 | 1.1.1 | |
| P.1.3 | packaging | | | | | 1.1.1 | |
| P.1.4 | mass | | | | 1.1.1 | | |

Table 9. Entities and attributes of the class structure 'piece goods' of the 'process domain' [source: author's illustration]

This example clarify the potential of the ontological framework. The framework offers a general and systemic knowledge to configure the best technical components and modules for the specific application due to the system technologies "robotics", "gripping technology", "pattern recognition" and "robot control and communication".

## 5. Conclusion

The paper presents an ontological approach to standardize robotic systems in logistic processes. Ontologies allow the systematic depiction of the technical systems in the procedural environment. Through their high level of abstraction, this chapter describes the conceptualization and elaboration of an ontological vocabulary for configuration process customized robotic architectures. The vocabulary allows the description of the relational structure for a dedicated reference process. It describes the relations among the procedural entities and the technical components. This ontology framework is the basis for the formation of modules and the configuration of the modules in robotics architectures.

The main goal provides a descriptive approach to the relationship between process and technology. Here, representations of conceptual ontologies were consulted. Due to the conceptual approach, the notation is on an abstract level, so that an automatic conclusion through formal ontologies is realistic. The representation of a dedicated solution space of possible technical configuration states of robotics system architectures is feasible, too.

In further research requirements, the development of formal ontologies in the context of this scope reduces the level of abstraction and enables the mechanical and automatic generation of ontologies.

In this way, interpretation and manipulation opportunities will be reduced and the interconnections of relationships between process and technology detailed. In this connection, formal ontologies allow the development of so-called architecture Configurator. They are based on the provided procedural and technical information and the possible ontological interrelationships. With this information, automatically development, including economic criteria, prioritizes configurations of robotic system architectures for dedicated reference processes. This approach can also serve as an appreciation of the nature of a 'Rapid Configuration Robotics' approach, which can digitally review prototyping activities such as technical feasibility and economic usefulness. The requirement for this type of IT-based configuration planning is shown by the RoboScan10 survey:

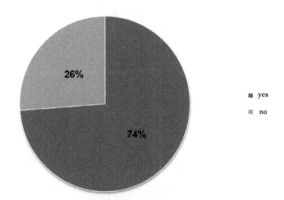

Fig. 11. Study RoboScan10: Answers about necessity for an IT-based system that plans the configuration of robotic systems [source: (Burwinkel, 2011)]

Fig. 11 shows the field of opinions in the context of RoboScan10 about the necessity for IT-based configuration planning of robotic systems. The question asked was: 'from a planning perspective: Could you envisage using an IT-based planning tool, which enables the configuration of both single robot systems and multi robot systems? 75% of all respondents could envisage the application of such tools.

## 6. References

Arnold, D. (2006). *Intralogistik : Potentiale, Perspektiven, Prognosen,* Springer, ISBN 978-3-540-29657-7, Berlin, Germany

Bateman, J. (1993): On the relationship between ontology construction and natural language: a socio-semiotic view. *International Journal of Human-Computer Studie,* Vol.43, No.5/6, pp.929-944, ISSN 1071-5819

Baumgarten, H. (2008). *Das Beste der Logistik : Innovationen, Strategien, Umsetzungen,* Springer, ISBN 978-3-540-78405-0, Berlin, Germany

Bunge, M. (1977). *Treatise on basic philosophy,* Reidel, ISBN 9027728399, Dordrecht, Netherlands

Burwinkel, M. & Pfeffermann, N. (2010). Die Zukunft der Robotik-Logistik liegt in der Modularisierung : Studienergebnisse "RoboScan'10". *Logistik für Unternehmen,* Vol.24, No.10, pp.21-23, ISSN 0930-7834

Chen, P. (1976). The entity-relationship model - toward a unified view of data. *ACM Trans. Database Syst,* Vol. 1, No. 4, pp. 9–36

Elger, J., & Haußener, C. (2009). Entwicklungen in der Automatisierungstechnik. In: *Internet der Dinge in der Intralogistik,* W. Günthner (Ed.), 23–27, Springer, ISBN 978-3-642-04895-1, Berlin, Germany

Erdmann, M. (2001). *Ontologien zur konzeptuellen Modellierung der Semantik von XML*, University Karlsruhe, ISBN 3-8311-2635-6, Karlsruhe, Germany

EUROP. (2009). Robotics Visions to 2020 And Beyond : The Strategic Research Agenda For Robotics in Europe, In: *European Robotics Technology Platform*, 20.06.2011, available from: www.robotics-platform.eu

Figgener, O. & Hompel, M. *Beitrag zur Prozessstandardisierung in der Intralogistik*. In: *Logistics Journal* (2007). ISSN 1860-5923 S. 1–12

Fritsch, D. & Wöltje, K. (2006). Roboter in der Intralogistik : Von der Speziallösung zum wirtschaftlichen Standardprodukt. *wt Werkstattstechnik Online*, Vol.96, No.9, pp. 623–630, ISSN 1436-4980

Gangemi, A., Guarino, N., Masolo, C., Oltramari, A. & Schneider, L. (2002). Sweetening Ontologies with DOLCE, In: *Knowledge Engineering and Knowledge Management: Ontologies and the Semantic Web*, Gómez-Pérez, A. & Benjamins, V. (Ed.),. pp.223–233, Springer, ISBN 978-3-540-44268-4, Berlin, Germany

Gómez-Pérez, A., Fernández-López, M., Corcho, O. (2004). *Ontological engineering: With examples from the areas of knowledge management, e-commerce and the semantic web*, Springer, ISBN 978-1-852-33551-9, London, England Göpfert, I. (2009). *Logistik der Zukunft - Logistics for the future*, Gabler, ISBN 978-3-834-91082-0, Wiesbaden, Germany

Gruber, T. (1993). A translation approach to portable ontology specifications. *Knowledge Acquisition*, Vol. 5, No.2, pp.199-220, ISSN 1042-8143

Gruninger, M. (2002). Ontology - Applications and Design. *Communication of the ACM*, Vol.45, No.2, pp.39–41, ISSN 00010782

Guarino, N. (1998). *Formal ontology in information systems: Proceedings of the first international conference (FOIS '98), June 6 - 8, Trento, Italy*, FOIS, ISBN 905-1-993-994, Amsterdam, Netherlands

Günthner, W. & HompelM. (2009.). *Internet der Dinge in der Intralogistik*, Springer, ISBN 978-3-642-04895-1, Berlin, Germany

Herb, M. (2006). Ontology Engineering mit OntoClean. In: *IPD University Karlsruhe*, 10.06.2011, available from: http://www.ipd.uni-karlsruhe.de/~oosem/S2D2/material/1-Herb.pdf

Kahraman, C., Cevik, S., Ates, N.Y. & Gulbay, M. (2007). Fuzzy multi-criteria evaluation of industrial robotic systems. *Computers & Industrial Engineering*, Vol.52, No.4, pp. 414-433, ISSN 0360-8352

Kastens, U. & Kleine Büning, H. (2008). *Modellierung : Grundlagen und formale Methoden*, Hanser, ISBN 978-3-446-41537-9, München, Germany

Kiencke, U. (1997). *Ereignisdiskrete Systeme: Modellierung und Steuerung verteilter Systeme*, Oldenbourg, ISBN 348-6-241-508, München, Germany

Kutz, O., Lutz, C., Wolter, F., Zakharyaschev, M. (2004). ε-connections of abstract description systems. *Artif. Intelligence*, Vol.156, No.1, pp.1-73, ISSN 0004-3702Neches, R., Fikes, R., Finin, T., Gruber, T., Patil, R., Senator, T. & Swartout, W. (1991). Enabling technology for knowledge sharing. *AI Mag*, Vol.12, No.3, pp.36–56

Scheid, W. (2010). Perspektiven zur Automatisierung in der Logistik : Teil 1 - Ansätze und Umfeld. *Hebezeuge Fördermittel - Fachzeitschrift für technische Logistik*, Vol.50, No.9, pp. 406–409, ISSN 0017-9442

Scheid, W. (2010): Perspektiven zur Automatisierung in der Logistik : Teil 2 - Praktische Umsetzung. *Hebezeuge Fördermittel - Fachzeitschrift für technische Logistik* Vol.50, No.10, pp. 482–483, ISSN 0017-9442

Schlenoff, C., Gruninger, M., Tissot, F., Valois, J. Lubell, J. & Lee, J. (1999). The Process Specification Language (PSL) Overview and Version 1.0 Specification, In: *www.mel.nist.gov/psl/*, 12.06.2011, available from: http://www.mel.nist.gov/msidlibrary/doc/nistir6459

Seidlmeier, H. (2002). *Prozessmodellierung mit ARIS® : Eine beispielorientierte Einführung für Studium und Praxis*, Vieweg, ISBN 352-8-058-048, Braunschweig, Germany

Siegert, H. (1996). *Robotik: Programmierung intelligenter Roboter*, Springer, ISBN 3540606653, Berlin, Germany

Staab, S. (2002). Wissensmanagement mit Ontologien und Metadaten. *Informatik-Spektrum*, Vol.25, No.3, pp.194–209, ISSN 0170-6012

Staud, J. (2006). *Geschäftsprozessanalyse : Ereignisgesteuerte Prozessketten und objektorientierte Geschäftsprozessmodellierung für betriebswirtschaftliche Standardsoftware*, Germany, ISBN 978-3-540-24510-0, Berlin, Germany

Steinmann, F. & Nejdl, W. (1999). Modellierung und Ontologie, In: Institut für Rechnergestützte Wissensverarbeitung, 25.05.2011, available from: www.kbs.uni-hannover.de/Arbeiten/Publikationen/1999/M%26O.pdf

Straube, F. & Rösch, F. (2008): *Logistik im produzierenden Gewerbe*, TU Berlin, ISBN 978-3-000-24165-9, Berlin, Germany

Stuckenschmidt, H. (2009). *Ontologien : Konzepte, Technologien und Anwendungen*, Springer, ISBN 978-3-540-79330-4, Berlin, Germany

Studer, R., Benjamins, V., Fensel, D. (1998). Knowledge engineering: Principles and methods. *Data & Knowledge Engineering*, Vol.25, , No.1-2, pp.161–197, ISSN 0169-023X

Suppa, M. & Hofschulte, J. (2010). Industrial Robotics. *at – Automatisierungstechnik*, Vol.58, No. 12., pp.663–664,

Swartout, W. , Tate, A. (1999). Ontologies. *IEEE Intelligent Systems and their Applications*, Vol.14, No.1, pp.18–19, ISSN 1094-7167

Tabeling, P. (2006): *Softwaresysteme und ihre Modellierung : Grundlagen, Methoden und Techniken*, Springer, ISBN 978-3-540-25828-5, Berlin, Germany

Westkämper, E., Verl,A. (2009). *Roboter in der Intralogistik : Aktuelle Trends - Neue Technologien - Moderne Anwendungen*, Verein zur Förderung produktionstechnischer Forschung, Stuttgart, Germany

Wiedemann, G. (2008). Ontologien und Ontology Engineering, In: Seminar 'Semantic Web Technologien', 12.06.2011, available from:     http://www.informatik.uni-leipzig.de/~loebe/teaching/2008ss-seweb/08v-ontengineering-gwiedemann.pdf

# 7

# Performance Evaluation of Fault-Tolerant Controllers in Robotic Manipulators

Claudio Urrea[1], John Kern[1,2] and Holman Ortiz[2]
*[1]Departamento de Ingeniería Eléctrica, DIE,*
*Universidad de Santiago de Chile, USACH, Santiago*
*[2]Escuela de Ingeniería Electrónica y Computación,*
*Universidad Iberoamericana de Ciencias y Tecnología, UNICIT, Santiago*
*Chile*

## 1. Introduction

Thanks to the incorporation of robotic systems, the development of industrial processes has generated a great increase in productivity, yield and product quality. Nevertheless, as far as technological advancement permits a greater automation level, system complexity also increases, with greater number of components, therefore rising the probability of failures or anomalous operation. This can result in operator's hazard, difficulties for users, economic losses, etc. Robotic automatic systems, even if helped in minimizing human operation in control and manual intervention tasks, haven't freed them from failure occurrences. Although such failures can´t be eliminated, they can be properly managed through an adequate control system, allowing to reduce degraded performance in industrial processes.

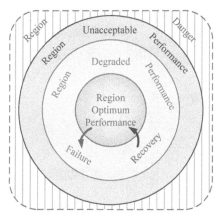

Fig. 1. Performance regions under failure occurrence

In figure 1 we see a scheme showing the different performance regions a given system can adopt when a failure occurs. If the system deviates to a degraded performance region in, presence of a failure, it can recover itself moving into an optimum performance region, or near to it. These systems are called fault tolerant systems and have become increasingly important for robot manipulators, especially those performing tasks in remote or hazardous environments, like outer space, underwater or nuclear environments.

In this chapter we will address the concept of fault tolerance applied to a robotic manipulator. We will consider the first three degrees of freedom of a redundant SCARA-type robot, which is intended to follow a Cartesian test trajectory composed by a combination of linear segments. We developed three fault-tolerant controllers by using classic control laws: *hyperbolic sine-cosine*, calculated torque and adaptive inertia. The essays for such controllers will be run in a simulation environment developed through MatLab/Simulink software. As a performance requirement for those controllers, we considered the application of a failure consisting in blocking one of the manipulator's actuators during trajectory execution. Finally, we present a performance evaluation for each one of the above mentioned fault-tolerant controllers, through joint and Cartesian errors, by means of graphics and rms rates.

## 2. Fault tolerant control

The concept of fault tolerant control (Zhang & Jiang, 2003) comes first from airplane fault tolerant control; although at scientific level it appears later, as a basic aim in the first congress of IFAC SAFEPROCESS 1991, with an especially stronger development since the beginning of 21th century. Fault tolerant control can be considered both under an active or passive approach, as seen in figure 2a. Passive tolerant control is based on the ability of feedback systems to compensate perturbations, changes in system dynamics and even system failures (Puig, Quevedo, Escobet, Morcego, & C., 2004). Passive tolerant control considers a robust design of the feedback control system in order to immunize it from some specific failures (Patton, 1997). Active tolerant control is centered in on-line failure, that is, the ability to identify the failing component, determine the kind of damage, its magnitude and moment of appearance and, from this information, to activate some mechanism for

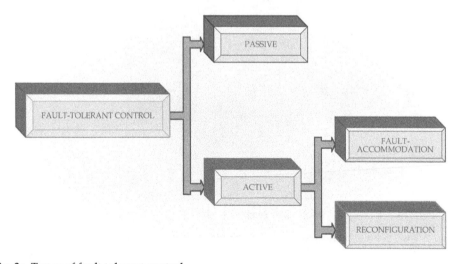

Fig. 2a. Types of fault tolerant control

rearrangement or control reconfiguration, even stopping the whole system, depending on the severity of the problem (Puig, Quevedo, Escobet, Morcego, & C., 2004).

Fault tolerant control systems (being of hybrid nature) consider the application of a series of techniques like: component and structure analysis; detection, isolation and quantification of failures; physical or virtual redundancy of sensors and/or actuators; integrated real-time supervision of all tasks performed by the fault tolerant control, as we can see in figure 2b (Blanke, Kinnaert, Lunze, & Staroswiecki, 2000).

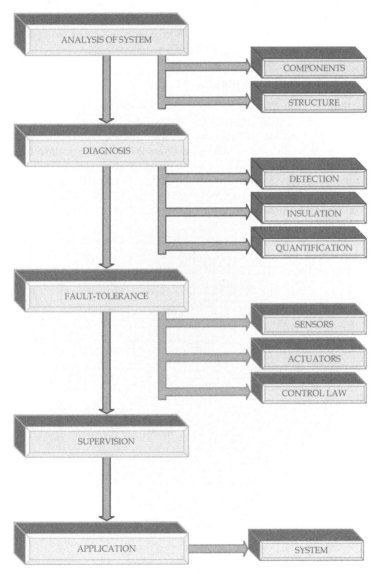

Fig. 2b. Stages included in the design of a fault tolerant control system

In the evaluation of fault tolerant controllers it is assumed that a robotic manipulator where a failure has arisen in one or more actuators, can be considered as an underactuated system, that is, a system with less actuators than the number of joints (El-Salam, El-Haweet, & and Pertew, 2005). Those underactuated systems present a greater degree of complexity compared with the simplicity of conventional robot control, being not so profoundly studied yet (Rubí, 2002). The advantages of underactuated systems have been recognized mainly because they are lighter and cheaper, with less energy consumption. Therefore, a great deal of concern is being focused on those underactuated robots (Xiujuan & Zhen, 2007). In figure 3 it is shown a diagram displaying the first three degrees of freedom of a SCARA type redundant manipulator, upon which essays will be conducted considering a failure in the second actuator, making the robot become an underactuated system.

## 3. SCARA-type redundant manipulator

For the evaluation of fault-tolerant controllers, we consider the first three degrees of freedom of a redundant  SCARA-type robotic manipulator, with a failure occurring in one of its actuators; such a system can be considered as an underactuated system, *i.e.*, with less actuators than the number of joints (Xiujuan & Zhen, 2007). Those underactuated systems have a greater complexity compared with the simplicity of conventional robots control, and they haven't been so deeply studied yet (Rubí, 2002). The advantages of underactuated systems have been remarked mainly because they are lighter and less expensive; also having less energy consumption, consequently an increasing level of attention is being paid to underactuated robots (Xiujuan & Zhen, 2007).

In figure 3 it is shown the scheme of a redundant SCARA-type robotic manipulator, and in figure 4 we can see a diagram showing the first three degrees of freedom of such manipulator, on which the essays will be carried on.

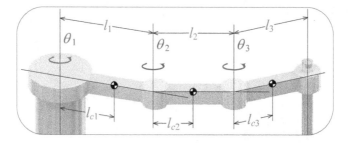

Fig. 3. Scheme of a SCARA-type redundant manipulator

The considered failure is the blocking of the second actuator, what makes this robot an underactuated system.

Having in mind the exposed manipulator, it is necessary to obtain its model; therefore we will consider that the dynamic model of a manipulator with $n$ joints can be expressed through equation (1):

$$\tau = M(q)\ddot{q} + C(q,\dot{q}) + G(q) + F(\dot{q}) \tag{1}$$

where:

$\tau$ :    Vector of generalized forces ($n{\times}1$ dimension).
$M$ :    Inertia matrix ($n{\times}n$ dimension).
$C$ :    Centrifugal and Coriolis forces vector ($n{\times}1$ dimension).
$q$ :    Components of joint position vector.
$\dot{q}$ :    Components of joint speed vector.
$G$ :    Gravity force vector ($n{\times}1$ dimension).
$\ddot{q}$ :    Joint acceleration vector ($n{\times}1$ dimension).
$F$ :    Friction forces vector ($n{\times}1$ dimension).

Under failure conditions in actuator number 2, that is, it's blocking, the component 2 of equation (1) becomes a constant.

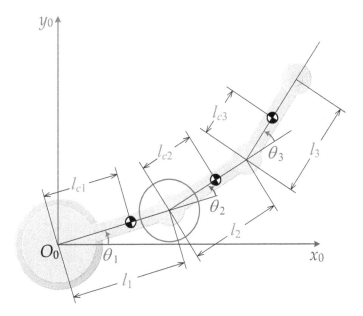

Fig. 4. Scheme of the three first DOF of a redundant SCARA-type robotic manipulator

## 4. Considered controllers

Considering the hybrid nature of fault tolerant control, it is proposed an active fault tolerant control having a different control law according to the status of the robotic manipulator, *i.e.*

normal or failing, with on-line sensing of possible failures and, in correspondence with this, reconfiguring the controller by selecting the most adequate control law (changing inputs and outputs).

Next, we will present a summary of the controllers considered for performance evaluation when a failure occurs in the second actuator of the previously described manipulator.

## 5. Fault tolerant controller: *hyperbolic sine* and *cosine*

This controller is based on the classic controller *hyperbolic sine-cosine* presented in (Barahona, Espinosa, & L., 2002), composed by a proportional part based on sine and *hyperbolic cosine* functions, a derivative part based on hyperbolic sine and gravity compensation, as shown in equation (2). The proposed fault tolerant control law includes two classic *hyperbolic sine-cosine* controllers that are "switched" to reconfigure the fault tolerant controller.

$$\tau = K_p \sinh(q_e)\cosh(q_e) - K_v \sinh(\dot{q}) + G(q) \tag{2}$$

$$q_e = q_d - q \tag{3}$$

According to equations (2) and (3):

$K_p$ : Proportional gain, diagonal definite positivematrix ($n \times n$ dimension).
$K_v$ : Derivative gain, diagonal definite positive matrix ($n \times n$ dimension).
$q_e$ : Joint position error vector ($n \times 1$ dimension).
$q_d$ : Desired joint position vector ($n \times 1$ dimension).

In (Barahona, Espinosa, & L., 2002) it is established that robotic manipulator's joint position error will tend asymptotically to zero as long as time approaches to infinite:

$$\lim_{t \to \infty} q_e \to 0 \tag{4}$$

This behavior is proved analyzing equation (5) and pointing that the only equilibrium point for the system is the origin (0,0).

$$\frac{d}{dt}\begin{bmatrix} q_e \\ \dot{q} \end{bmatrix} = \begin{bmatrix} a_1 \\ a_2 \end{bmatrix}$$

$$a_1 = -\dot{q} \tag{5}$$

$$a_2 = M(q)^{-1}\left(K_p \sinh(q_e)\cosh(q_e) - K_v \sinh(\dot{q}) - C(q,\dot{q})\dot{q}\right)$$

## 6. Fault tolerant controller: Computed torque

Another active fault tolerant controller analyzed here uses a control law by computed torque, consisting in the application of a torque in order to compensate the centrifugal, Coriolis, gravity and friction effects, as shown in equation (6).

$$\tau = \hat{M}(q)\left(\ddot{q}_d + K_v\dot{q}_e + K_p q_e\right) + \hat{C}(q,\dot{q}) + \hat{G}(q) + \hat{F}(\dot{q}) \tag{6}$$

where:

$\hat{M}$ : Estimation of inertia matrix ($n \times n$ dimension).

$\hat{C}$ :    Estimation of centrifugal and Coriolis forces vector ($n\times1$dimension).
$\hat{G}$ :    Estimation of gravity force vector ($n\times1$dimension).
$\hat{F}$ :    Estimation of friction forces vector ($n\times1$dimension).

$$K_v = \begin{bmatrix} K_{v1} & & & \\ & K_{v2} & & \\ & & \ddots & \\ & & & K_{vn} \end{bmatrix} \tag{7}$$

$K_v$ :    Diagonal definite positive matrix ($n\times n$ dimension).

$$K_p = \begin{bmatrix} K_{p1} & & & \\ & K_{p2} & & \\ & & \ddots & \\ & & & K_{pn} \end{bmatrix} \tag{8}$$

$K_p$ :    Diagonal definite positive matrix ($n\times n$ dimension).
$\ddot{q}_d$ :    Desired joint acceleration vector ($n\times1$ dimension).

$$\dot{q}_e = \dot{q}_d - \dot{q} \tag{9}$$

$\dot{q}_e$ :    Joint speed error vector ($n\times1$dimension).
If estimation errors are little, joint errors near to a linear equation, as shown in equation (10).

$$\ddot{q}_e + K_v\dot{q}_e + K_pq_e \approx 0 \tag{10}$$

## 7. Fault tolerant controller: Adaptive inertia

The fault tolerant control under examination is based on an adaptive control law, namely: adaptive inertia (Lewis, Dawson, & Abdallah, 2004), (Siciliano & Khatib, 2008), for what it is necessary to consider the manipulator dynamic model in the form expressed in equation (11). The term corresponding to centrifugal and Coriolis forces is expressed through a matrix $V_m$.

$$\tau = M(q)\ddot{q} + V_m(q,\dot{q})\dot{q} + G(q) + F(\dot{q}) \tag{11}$$

In this case, we define an auxiliary error signal $r$ and its derivative $\dot{r}$, as shown in equations (12) and (13), respectively:

$$r = \Lambda q_e + \dot{q}_e \tag{12}$$

$$\dot{r} = \Lambda \dot{q}_e + \ddot{q}_e \tag{13}$$

where:
$\Lambda$ :    Diagonal definite positive matrix ($n\times n$ dimension).

$$\Lambda = \begin{bmatrix} \lambda_1 & & & \\ & \lambda_2 & & \\ & & \ddots & \\ & & & \lambda_n \end{bmatrix} \tag{14}$$

When replacing equations (3), (9), (12) and (13) into expression (11), we obtain:

$$\tau = M(q)(\ddot{q}_d + \Lambda\dot{q}_e) + V_m(q,\dot{q})(\dot{q}_d + \Lambda q_e) + G(q) + F(\dot{q}) - M(q)\dot{r} - V_m(q,\dot{q})r \tag{15}$$

And making the following matching:

$$Y(\cdot)\phi = M(q)(\ddot{q}_d + \Lambda\dot{q}_e) + V_m(q,\dot{q})(\dot{q}_d + \Lambda q_e) + G(q) + F(\dot{q}) \tag{16}$$

where:

$$Y(q,\dot{q},q_d,\dot{q}_d,\ddot{q}_d) = \begin{bmatrix} Y_{11} & Y_{12} & \cdots & Y_{1n} \\ Y_{21} & Y_{22} & \cdots & Y_{2n} \\ \vdots & \vdots & \ddots & \vdots \\ Y_{n1} & Y_{n2} & \cdots & Y_{nn} \end{bmatrix} \tag{17}$$

$Y(\cdot)$ : Regression matrix ($n \times n$ dimension).
$\phi$ : Parameter vector ($n \times 1$ dimension).
With these relationships, expression (15) can be rewritten in the following way:

$$\tau = Y(\cdot)\phi - M(q)\dot{r} - V_m(q,\dot{q})r \ Y(\cdot) \tag{18}$$

And the control torque is expressed through equation (19):

$$\tau = Y(\cdot)\hat{\phi} + K_v r \tag{19}$$

where:
$\hat{\phi}$ : Parameter estimation vector ($n \times 1$ dimension).
$K_v$: Diagonal definite positive matrix ($n \times n$ dimension).
The adaptive control updating rule can be expressed by:

$$\dot{\hat{\phi}} = -\dot{\tilde{\phi}} = \Gamma Y^T(\cdot)r \tag{20}$$

where:
$\Gamma$: Diagonal definite positive matrix ($n \times n$ dimension).

## 8. Fault tolerant control simulator

The three above mentioned control laws, along with the dynamic model of the redundant SCARA-type manipulator considering the first three degrees of freedom (Addendum A), are run under the simulation structure shown in figure 5, where we can see the hybrid nature of this kind of controller.

In Addendum B we show the set of parameter values employed in the manipulator dynamic model, and the gains considered for each kind of fault tolerant controller.

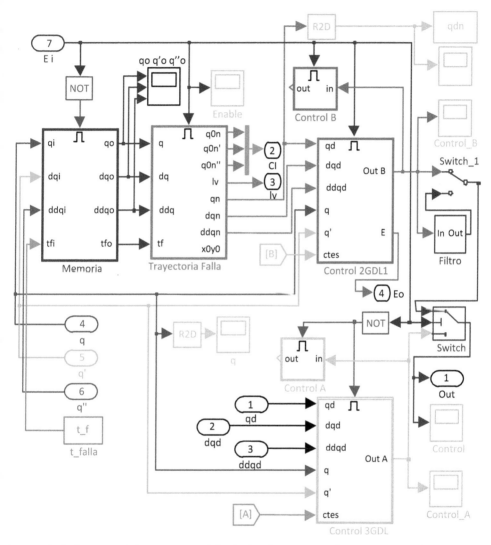

Fig. 5. Block diagram of the structure of the fault tolerant controller used to test the above mentioned control laws

## 9. Results

After establishing the control laws being utilized, we determine the trajectory to be entered in the control system to carry out the corresponding performance tests of fault tolerant control algorithms. This trajectory is displayed in figure 6.

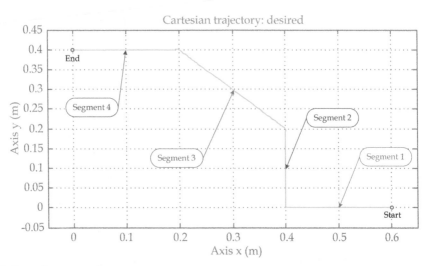

Fig. 6. Cartesian test trajectory

Figures 7 and 8a show the curves corresponding to the differences between desired and real joint trajectories, and between desired and real Cartesian trajectories, respectively, all this under *hyperbolic sine-cosine* fault tolerant control when there is a failure in actuator 2 at 0.5 sec from initiating movement.

Where:

$e_{q1}$ :    Joint trajectory error, joint 1.
$e_{q2}$ :    Joint trajectory error, joint 2.
$e_{q3}$ :    Joint trajectory error, joint 3.
$e_x$  :    Cartesian trajectory error, $x$ axis.
$e_y$  :    Cartesian trajectory error, $y$ axis.

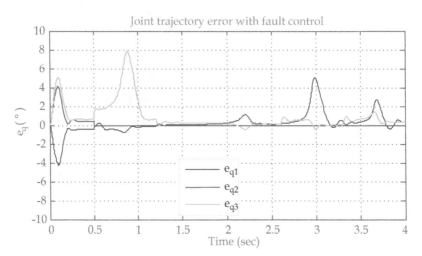

Fig. 7. Joint trajectory error with fault control using *hyperbolic sine-cosine* controller

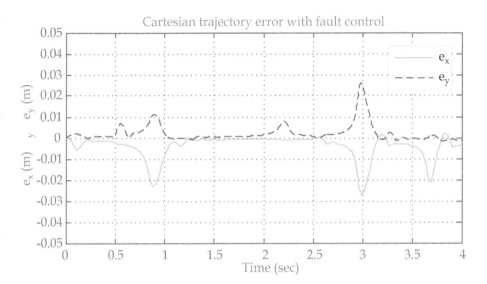

Fig. 8a. Cartesian trajectory error with fault control using *hyperbolic sine-cosine* controller

The performance of fault tolerant controller by computed torque is shown in figures 8b and 9, displaying the curves for joint and Cartesian errors under the same failure conditions than the previous case.

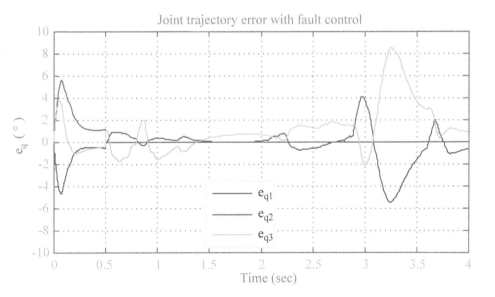

Fig. 8b. Joint trajectory error with fault control using computed torque controller

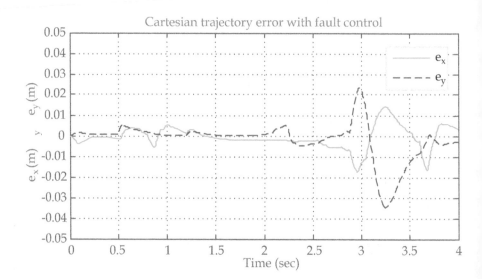

Fig. 9. Cartesian trajectory error with fault control using computed torque controller

In figures 10 and 11 we can see charts displaying respectively joint and Cartesian errors corresponding to the performance of fault tolerant controller by adaptive inertia, under the same failure conditions imposed to the previous controllers.

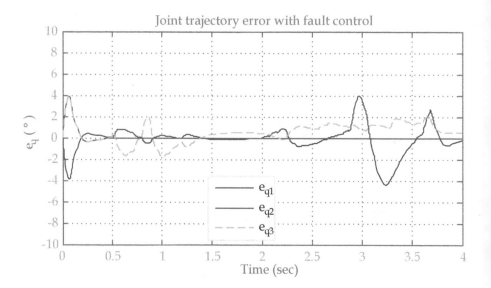

Fig. 10. Joint trajectory error with fault control using adaptive inertia controller

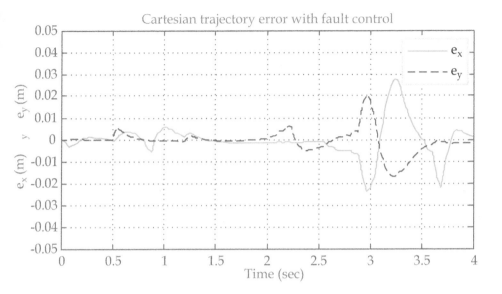

Fig. 11. Cartesian trajectory error with fault control using adaptive inertia controller

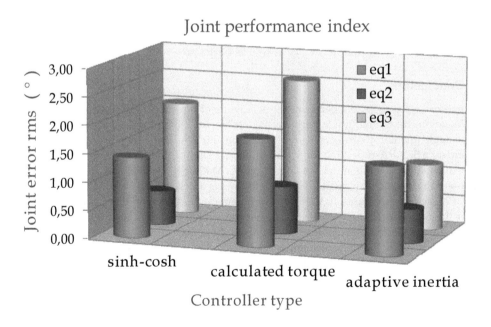

Fig. 12. Performance index corresponding to joint trajectory

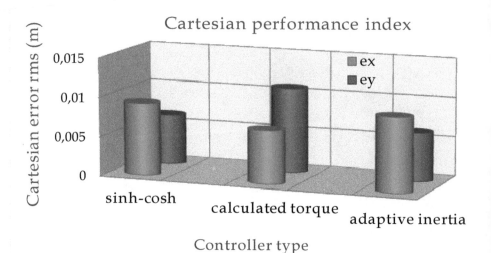

Fig. 13. Performance index corresponding to Cartesian trajectory

Finally, in figures 12 and 13 it is shown a performance summarization of the analyzed fault tolerant controllers in terms of joint and Cartesian *mean square root* errors, accordingly to equation (21)

$$rms = \sqrt{\frac{1}{n}\sum_{i=1}^{n} e_i^2} \tag{21}$$

Where $e_i$ represents articular trajectory as well as Cartesian errors.

## 10. Conclusions

In this work we presented a performance evaluation of three fault tolerant controllers based on classic control techniques: hyperbolic sine-cosine, calculated torque and adaptive inertia. Those fault tolerant controllers were applied on the first three degrees of freedom of a redundant SCARA-type robotic manipulator. The different system stages were implemented in a simulator developed using MatLab/Simulink *software*, allowing to represent the robotic manipulator behavior following a desired trajectory, when blocking of one of its actuators occurs. In this way we obtained the corresponding simulation curves. From the obtained results, we observed that the adaptive inertia fault tolerant controller have errors with less severe maximums than the other controllers, resulting in more homogeneous manipulator movements. We noticed that greater errors were produced with the calculated torque fault tolerant controller, both for maximums and *rms*. Consequently, the best performance is obtained when using the adaptive inertia controller, as shown in figures 14 and 15. It is also remarkable that the hyperbolic sine-cosine fault tolerant controller have a lesser implementation complexity, since it does not require the second derivative of joint position. This can be a decisive factor in the case of not having high performance processors.

## 11. Further developments

Thanks to the development of this work, from the implemented simulation tools and the obtained results, fault tolerant control systems essays are being currently carried out, in order to apply them to actual robotic systems, with and without link redundancy, like the SCARA-type robots shown in figure 14 and figure 15, respectively.

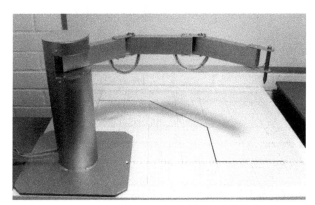

Fig. 14. SCARA-type redundant robot, DIE-USACH

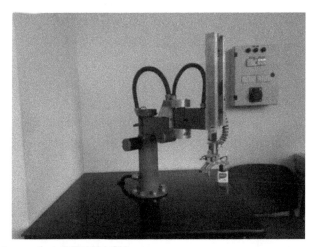

Fig. 15. SCARA-type robot, DIE-USACH

## 12. Addendum A: Manipulator's dynamic model

The manipulator's dynamic model is given by equations a1 to a14.

$$\mathbf{M} = \begin{bmatrix} M_{11} & M_{12} & M_{13} \\ M_{21} & M_{22} & M_{23} \\ M_{31} & M_{32} & M_{33} \end{bmatrix} \tag{a1}$$

$$M_{11} = I_{1zz} + I_{2zz} + I_{3zz} + m_1 l_{c1}^2 + m_2 \left( l_1^2 + l_{c2}^2 \right) + m_2 2 l_1 l_{c2} \cos \theta_2 + ...$$
$$m_3 \left( l_1^2 + l_2^2 + l_{c3}^2 + 2 l_1 l_2 \cos \theta_2 + 2 l_2 l_{c3} \cos \theta_3 + 2 l_1 l_{c3} \cos \left( \theta_2 + \theta_3 \right) \right)$$

(a2)

$$M_{21} = M_{12} = I_{2zz} + I_{3zz} + m_2 \left( l_{c2}^2 + l_1 l_{c2} \cos \theta_2 \right) + ...$$
$$m_3 \left( l_2^2 + l_{c3}^2 + l_1 l_2 \cos \theta_2 + 2 l_2 l_{c3} \cos \theta_3 + l_1 l_{c3} \cos \left( \theta_2 + \theta_3 \right) \right)$$

(a3)

$$M_{31} = M_{13} = I_{3zz} + m_3 \left( l_{c3}^2 + l_2 l_{c3} \cos \theta_3 \right) + m_3 l_1 l_{c3} \cos \left( \theta_2 + \theta_3 \right)$$

(a4)

$$M_{22} = I_{2zz} + I_{3zz} + m_2 l_{c2}^2 + m_3 \left( l_2^2 + l_{c3}^2 + 2 l_2 l_{c3} \cos \theta_3 \right)$$

(a5)

$$M_{32} = M_{23} = I_{3zz} + m_3 \left( l_{c3}^2 + l_2 l_{c3} \cos \theta_3 \right)$$

(a6)

$$M_{33} = I_{3zz} + m_3 l_{c3}^2$$

(a7)

$$\mathbf{C} = \begin{bmatrix} C_{11} & C_{21} & C_{31} \end{bmatrix}^T$$

(a8)

$$C_{11} = -2 l_1 \left( m_2 l_{c2} \sin \theta_2 + m_3 l_2 \sin \theta_2 \right) \dot{\theta}_1 \dot{\theta}_2 - 2 l_1 m_3 l_{c3} \sin \left( \theta_2 + \theta_3 \right) \dot{\theta}_1 \dot{\theta}_2 + ...$$
$$- m_2 l_1 l_{c2} \sin \theta_2 \cdot \dot{\theta}_2^2 + m_3 \left( l_1 l_2 \sin \theta_2 + l_1 l_{c3} \sin \left( \theta_2 + \theta_3 \right) \right) \dot{\theta}_2^2 + ...$$
$$- 2 l_{c3} m_3 \left( l_2 \sin \theta_3 + l_1 \sin \left( \theta_2 + \theta_3 \right) \right) \dot{\theta}_1 \dot{\theta}_3 + ...$$
$$- 2 m_3 l_{c3} \left( l_2 \sin \theta_3 + l_1 \sin \left( \theta_2 + \theta_3 \right) \right) \dot{\theta}_2 \dot{\theta}_3 + ...$$
$$m_3 \left( -l_2 l_{c3} \sin \theta_3 - l_1 l_{c3} \sin \left( \theta_2 + \theta_3 \right) \right) \dot{\theta}_3^2$$

(a9)

$$C_{21} = m_3 \left( l_1 l_2 \sin \theta_2 + l_1 l_{c3} \sin \left( \theta_2 + \theta_3 \right) \right) \dot{\theta}_1^2 + m_2 l_1 l_{c2} \sin \theta_2 \cdot \dot{\theta}_1^2 + ...$$
$$- 2 m_3 l_2 l_{c3} \sin \theta_3 \cdot \dot{\theta}_1 \dot{\theta}_3 - 2 m_3 l_2 l_{c3} \sin \theta_3 \cdot \dot{\theta}_2 \dot{\theta}_3 - m_3 l_2 l_{c3} \sin \theta_3 \cdot \dot{\theta}_3^2$$

(a10)

$$C_{31} = m_3 \left( l_2 l_{c3} \sin \theta_3 + l_1 l_{c3} \sin \left( \theta_2 + \theta_3 \right) \right) \dot{\theta}_1^2 + ...$$
$$2 m_3 l_2 l_{c3} \sin \theta_3 \cdot \dot{\theta}_1 \dot{\theta}_2 + m_3 l_2 l_{c3} \sin \theta_3 \cdot \dot{\theta}_2^2$$

(a11)

$$C_{31} = m_3 \left( l_2 l_{c3} \sin \theta_3 + l_1 l_{c3} \sin \left( \theta_2 + \theta_3 \right) \right) \dot{\theta}_1^2 + 2 m_3 l_2 l_{c3} \sin \theta_3 \cdot \dot{\theta}_1 \dot{\theta}_2 + ...$$
$$m_3 l_2 l_{c3} \sin \theta_3 \cdot \dot{\theta}_2^2$$

(a12)

$$\mathbf{G} = \begin{bmatrix} 0 & 0 & 0 \end{bmatrix}^T$$

(a13)

$$\mathbf{F} = \begin{bmatrix} F_{11} & F_{21} & F_{31} \end{bmatrix}^T$$

(a14)

where:

$m_1$ : First link mass.

$m_2$ : Second link mass.

$m_3$ : Third link mass.
$l_1$ : First link length.
$l_2$ : Second link length.
$l_3$ : Third link length.
$l_{c1}$ : Length from 1st link origin to its centroid.
$l_{c2}$ : Length from 2nd link origin to its centroid.
$l_{c3}$ : Length from 3rd link origin to its centroid.
$I_{1zz}$ : 1st link inertial momentum with respect to the first $z$ axis of its joint.
$I_{2zz}$ : 2nd link inertial momentum with respect to the first $z$ axis of its joint.
$I_{3zz}$ : 3rd link inertial momentum with respect to the first $z$ axis of its joint.

## 13. Addendum B: Considered parameter values

Parameter values considered for the manipulator as well as controller gains values are shown in tables B1 and B2, respectively.

| | Link 1 | | Link 2 | | Link 3 | Units |
|---|---|---|---|---|---|---|
| $l_1$ | = 0.2 | $l_2$ | = 0.2 | $l_3$ | = 0.2 | $[\mathrm{m}]$ |
| $l_{c1}$ | = 0.0229 | $l_{c2}$ | = 0.0229 | $l_{c3}$ | = 0.0983 | $[\mathrm{m}]$ |
| $m_1$ | = 2.0458 | $m_2$ | = 2.0458 | $m_3$ | = 6.5225 | $[\mathrm{kg}]$ |
| $I_{1zz}$ | = 0.0116 | $I_{2zz}$ | = 0.0116 | $I_{3zz}$ | = 0.1213 | $\left[\mathrm{kg \cdot m^2}\right]$ |
| $F_{v1}$ | = 0.025 | $F_{v2}$ | = 0.025 | $F_{v3}$ | = 0.025 | $\left[\dfrac{\mathrm{N \cdot m \cdot s}}{\mathrm{rad}}\right]$ |
| $F_{eca1}$ | = 0.05 | $F_{eca2}$ | = 0.05 | $F_{eca3}$ | = 0.05 | $[\mathrm{N \cdot m}]$ |
| $F_{ecb1}$ | = -0.05 | $F_{ecb2}$ | = -0.05 | $F_{ecb3}$ | = -0.05 | $[\mathrm{N \cdot m}]$ |

Table B1. Considered parameters for the manipulator

| | Controller Type | | |
|---|---|---|---|
| Constants | Hyperbolic Sine-Cosine | Computed Torque | Adaptive Inertia |
| $K_{p1}, K_{p2}, K_{p3}$ | 400, 300, 200 | 800, 800, 800 | — |
| $K_{v1}, K_{v2}, K_{v3}$ | 3, 2, 1 | 140, 140, 140 | 20, 20, 20 |
| $\lambda_1, \lambda_2, \lambda_3$ | — | — | 8, 8, 8 |
| $\gamma_1, \gamma_2, \gamma_3$ | — | — | 0.1, 0.1, 0.1 |

Table B2. Controller gains

## 14. References

Barahona, J., Espinosa & L., C.F., 2002. Evaluación Experimental de Controladores de Posición tipo Saturados para Robot Manipuladores. In *Congreso Nacional de Electrónica, Centro de Convenciones William o Jenkins*. Puebla. México, 2002.

Blanke, M., Kinnaert, M., Lunze, J. & Staroswiecki, M., 2000. What is Fault-Tolerant Control. In *IFAC Symposium on Fault Detection, Supervision and Safety for Technical Process - SAFEPROCESS 2000*. Budapest, 2000. Springer-Verlag Berlin Heidelberg.

El-Salam, A., El-Haweet, W. & and Pertew, A., 2005. Fault Tolerant Kinematic Controller Design for Underactuated Robot Manipulators. In *The Automatic Control and Systems Engineering Conference*. Cairo, 2005.

Lewis, F., Dawson, D. & Abdallah, C., 2004. *Robot Manipulator Control Theory and Practice*. New York: Marcel Dekker, Inc.

Patton, R.J., 1997. Fault-Tolerant Control: The 1997 Situation. In *Proc. IFAC Symposium Safeprocess*. Kingston Upon Hull, 1997.

Puig, V. et al., 2004. Control Tolerante a Fallos (Parte I): Fundamentos y Diagnóstico de Fallos. *Revista Iberoamericana de Automática e Informática Industrial*, 1(1), pp.15-31.

Rubí, J., 2002. *Cinemática, Dinámica y Control de Robots Redundantes y Robots Subactuados*. Tesis Doctoral. San Sebastián, España: Universidad de Navarra.

Siciliano, B. & Khatib, O., 2008. *Handbook of Robotics*. Berlin, Heidelberg: Springer-Verlag.

Xiujuan, D. & Zhen, L., 2007. *Underactuated Robot Dynamic Modeling and Control Based on Embedding Model*. In *12th IFToMM World Congress*. Besançon. France, 2007.

Zhang, Y. & Jiang, J., 2003. Bibliographical Review on Reconfigurable Fault-Tolerant Control Systems. In *Proceedings IFAC SAFEPROCESS*. Washington, D.C., USA, 2003.

# 8

# An Approach to Distributed Component-Based Software for Robotics*

A. C. Domínguez-Brito, J. Cabrera-Gámez,
J. D. Hernández-Sosa, J. Isern-González and E. Fernández-Perdomo
*Instituto Universitario SIANI & the Departamento de Informática y Sistemas,*
*Universidad de Las Palmas de Gran Canaria*
*Spain*

## 1. Introduction

Programming robotic systems is not an easy task, even developing software for simple systems may be difficult, or at least cumbersome and error prone. Those systems are usually multi-threaded and multi-process, so synchronization problems associated with processes and threads must be faced. In addition distributed systems in network environments are also very common, and coordination between processes and threads in different machines increases programming complexity, specially if network environments are not completely reliable like a wireless network. Hardware abstraction is another question to take into account, uncommon hardware for an ordinary computer user is found in robotics systems, sensors and actuators with APIs (Applications Programming Interfaces) in occasions not very stable from version to version, and many times not well supported on the most common operating systems. Besides, it is not rare that even sensors or actuators with the same functionality (i.e. range sensors, cameras, etc.) are endowed with APIs with very different semantics. Moreover, many robotic systems must operate in hard real time conditions in order to warrant system and operation integrity, so it is necessary that the software behaves obeying strictly specific response times, deadlines and high frequencies of operation. Software integration and portability are also important problems in those systems, since many times only in one of them we may find a variety of machines, operating systems, drivers and libraries which we have to cope to. Last but not least, we want robotic systems to behave "autonomous" and "intelligently", and to carry out complex tasks like mapping a building, navigating safely in a cluttered, dynamic and crowded environment or driving a car safely in a motorway, to name a few.

Despite there is no established standard methodology or solution to the situation described in the previous paragraph, in the last ten years many approaches have blossomed in the robotics community in order to tackle with it. In fact, many software engineering techniques and experiences coming from other areas of computer science are being applied to the specific area of robotic control software. A review of the state of the art of software engineering applied

*This work has been partially supported by the Research Project *TIN2008-06068* funded by the Ministerio de Ciencia y Educación, Gobierno de España, Spain, and by the Research Project *ProID20100062* funded by the Agencia Canaria de Investigación, Innovación y Sociedad de la Información (ACIISI), Gobierno de Canarias, Spain, and by the European Union's FEDER funds.

specifically to robotics can be found in [Brugali (2007)]. Many of the approaches that have come up in these last years, albeit different, are either based completely, or follow or share to a certain extend some of the fundamentals ideas of the CBSE (Component Based Software Engineering) [George T. Heineman & William T. Councill (2001)] paradigm as a principle of design for robotic software.

Some of the significant approaches freely available within the robotics community based on the CBSE paradigm are $G^{en}oM/BIP$ [Mallet et al. (2010)][Basu et al. (2006)][Bensalem et al. (2009)], Smartsoft [Schlegel et al. (2009)], OROCOS [The Orocos Project (2011)], project ORCA [Brooks et al. (2005)], OpenRTM-aist [Ando et al. (2008)] and Willow Garage's ROS project [ROS: Robot Operating System (2011)]. All these approaches, in general incompatible among them, use many similar abstractions in order to build robotic software out of a set of software components. Each of these approaches usually solve or deal with many of the mentioned problems faced when programming robotic systems (hard real-time operation, distributed, multithread and multiprocess programming, hardware abstractions, portability, etc.), and using any of them implies to get used to its own methodology, abstractions and software tools to develop robotic software. Our group have also developed an approach to tackle with the problem of programming robotic systems. It is some years already that we have been using a CBSE C++ distributed framework designed and developed in our laboratory, termed CoolBOT [Antonio C. Domínguez-Brito et al. (2007)], which is also aimed at easing software development for robotic systems. Along several years of use acquiring experience programming mobile robotic systems, we have ended up integrating in CoolBOT some new developments in order to improve their use and operation. Those new improvements have been focused mainly in two main questions, namely: transparent distributed computation, and "deeper" interface decoupling; in next sections they will be presented more deeply. In order to do so this paper is organized as follows. In section 2 we will introduce briefly an overview about CoolBOT. Next, we will pass to focus on each one of the mentioned topics respectively, in sections 4 and 5. The last section is devoted to presenting the conclusions of this work.

## 2. CoolBOT. Overview

CoolBOT [Antonio C. Domínguez-Brito et al. (2007)] is a C++ component oriented programming framework aimed to robotics, developed at our laboratory some years ago [Domínguez-Brito et al. (2004)], which is normally used to develop the control software for the mobile robots we have available at our institution. It is a programming framework that follows the CBSE paradigm for software development. The key concept in the CBSE paradigm is the concept of *software component* which is a unit of integration, composition and software reuse [Brugali & Scandurra (2009)][Brugali & Shakhimardanov (2010)]. Complex systems might be composed of several ready-to-use components. Ideally, interconnecting available components out of a repository of components previously developed, we can program a complete system. Thus, it should be only necessary a graphical interface or alike, to set up a system. Hence, being CoolBOT CBSE oriented, it also makes use of this central concept to build software systems.

Fig. 1 gives a global view of a typical system developed using CoolBOT. As we can observe, there are five CoolBOT *components* and two CoolBOT *views*, forming all them four CoolBOT *integrations* involving three different machines sharing a computer network. In addition, hosted by one of the machines, there is a non CoolBOT application which uses a CoolBOT *probe* to interact with one of the components of the system. Thus, in CoolBOT we can find three types of software components: *components*, *views* and *probes*. All these three types

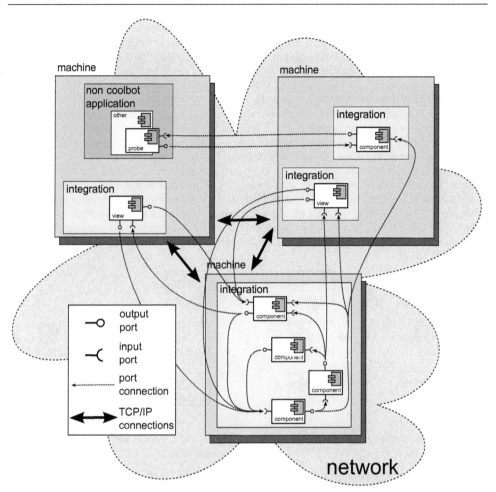

Fig. 1. Diagram of the elements and interconnections of a system designed with CoolBOT.

are software components in the whole sense, since we can compose them indistinctly and arbitrarily without changing their implementation to built up a given system. The main difference among them is that *views* and *probes* are "light-weight" software components in relation to CoolBOT *components*. *Views* are software components which implement graphical control and monitoring interfaces for CoolBOT systems, which are completely decoupled from them. On the other side, *probes* mainly allow to implement decoupled interfaces for interoperation of CoolBOT systems with non CoolBOT applications, as depicted in the figure. Both will be explained in more detail in section 5.

In CoolBOT, systems are made of CoolBOT *components* (components for short). A component is an *active entity*, in the sense that it has its own unit of execution. It also presents a clear separation between its external interface and its internals. Components intercommunicate among them only through their external interfaces which are formed by *input and output ports*. When connected, they form *port connections*, as depicted on Fig. 1. Through them, components interchange discrete units of information termed *port packets*. *Views* and *probes* have a similar external interface of input and output ports, hence, they can also be interconnected among them and with components using port connections. The functionality of a whole system comes up from the interaction through port connections among all the components integrating the system, including views and probes.

## 2.1 Port connections, ports and port packets

CoolBOT components interact among them using *port connections*. A *port connection* is defined by an output port and an input port. Port connections are established dynamically in runtime, and they are unidirectional (from output to input port), and follow a *publish/subscribe* paradigm of communication [Brugali & Shakhimardanov (2010)]. In that way, we can have multiple subscribers for the same output port, as shown in Fig. 2, and multiple publishers feeding the same input port, illustrated in Fig. 3. Note that input and output ports are decoupled in the sense that component publishers do not know necessarily who is receiving what they are publishing through their output ports. The contrary is also true, component subscribers do not necessarily know who is publishing the data which are reaching them through their input ports. Data are sent through port connections in discrete units called *port packets*.

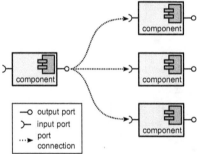

Fig. 2. Port connections: one publisher, many subscribers.

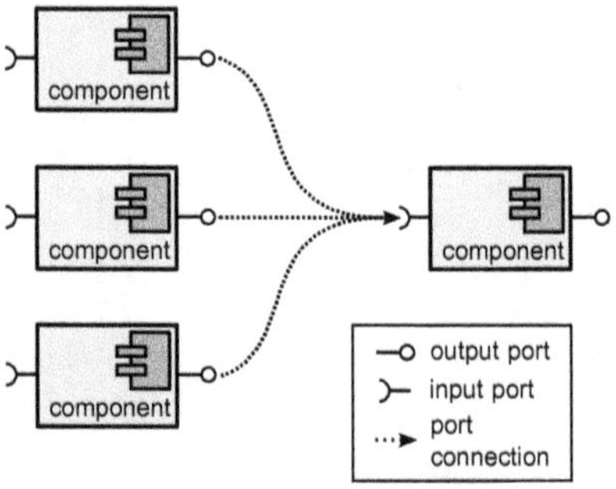

Fig. 3. Port connections: many publishers, one subscriber.

To establish a port connection the ports involved should be compatible, i.e., they must have compatible types, and should transport the same types of port packets. In particular, when defining an output or input port we have to specify three aspects for them, namely:

1. **An identifier:** This is an identifier or name for the port. It has to be unique at component (or *view* or *probe*) scope. The port identifier is what we use to refer to a specific port when establishing/de-establishing port connections.

2. **A port type:** This is the type of the port. There are several typologies available for input and output ports, and depending on how we combine them, we can establish a different model of communication for each connection. The typologies of the input and output ports involved in a connection determine the pattern and semantics of the communication through it, following the same philosophy that the communications patterns of Smartsoft [Schlegel et al. (2009)], and the interfaces for component communication available in OROCOS [*The Orocos Project* (2011)]. On Tables 1 and 2 we can see all the types of connections we have available in CoolBOT right now, we will elaborate on this later.

3. **Port packet types:** Those are the types of port packets accepted by the output or input port. Most input and output port types only accept one type of port packet through them, although we have also some of them that accept a set of different port packet types.

Bear in mind that in CoolBOT port connections are established dynamically, but the definition of each input and output port for each software component, whether a component, a view or a probe, is static. Thus, for a given component we define statically its external interface of input and output ports, each one with its identifier, port type and accepted port packet types. In opposition, port connections are established at runtime. Only a compatible pair of output and input ports can form a port connection, and we say that they are compatible when two conditions fulfill: first, they have compatible port types (the compatible combinations are shown in Tables 1 and 2); and second, the port packets the pair of port accepts also match.

As commented, Tables 1 and 2 show all the possible types and combinations of output and input ports available in CoolBOT. As we can observe we have two groups of types of port connections depending on the types of the output and input ports we connect, namely:

- *Active Publisher/Passive Subscriber (AP/PS)* **connections.** In this kind of connections we say the publisher is the *active* part of the communication through the connection, since it is the publisher (the sender) who invest more computational resources in the communication. More specifically, in these connections, there is a buffer (a cache, a memory) in the input port where incoming port packets get stored, when they get to the subscriber (the receiver) end. We say the publisher is active, because the copy of port packets in the input port buffers is made by the publisher's threads of execution. Those memories get signaled for the subscribers to access them at their own pace. Evidently, if the output port has several subscribers, the publisher has to make a copy for all of them, so the computational cost for copies increases and this cost is afforded by the publisher. Table 1 enumerates all the available types of port connections following this model of communication.

- *Passive Publisher/Active Subscriber (PP/AS)* **connections.** Those connections follow a model of communication where the subscriber plays the *active* role in the communication, in the sense that in opposition to the previous ones, the subscriber is the part of the communication which invests more computational resources. In this type of connections, we have buffers at both ends of the port connection. When the publisher sends (publishes) a port packet through the connection, it gets stores in a buffer in the output port, and the input port gets signaled, in order to notify the subscriber that there are new data on the connection. Note that the publisher does not copy the port packet on the subscriber's input port buffer. It is the subscriber the one who copies the port packet on its input port buffer when it access its input port in order to get fresh port packets, stored at the other end of the port connection. In this way, when we have several subscribers and one publisher the computational cost of copies is afforded by each one of the subscribers separately.

Apart from the computational cost of using AP/PS connections versus PP/AS connections, there is another important aspect to take into account when using any of them. PP/AS connections are *persistent* in the sense that, as explained in Table 2 the last data (port packet) sent by the publisher through the output port is stored there, in the output port buffer, so subscribers which are connected to this output port once the last port packet was sent, can access those data posteriorly. On the contrary AP/PS connections are not persistent, because port packets are stored on the subscriber end, so packets, once they have been sent, are not available for new subscribers, because port packets only gets to those subscribers who are connected to the output port right at the moment of sending them.

Other important aspect to take into account with respect to port connections is that they follow an asynchronous model of communications. Note that they are unidirectional, port packets go from the publisher's output port to the subscriber's input port, the publisher sends port

| Active Publisher/Passive Subscriber (AP/PS) | | |
|---|---|---|
| Output Port | Input Port | Port Connection Type |
| *tick* | *tick* | *tick* connections: those connections do not transport any port packet, they only communicate the occurrence of an event. |
| *generic* | *last* | *last* connections: the input port stores always the *last* port packet sent through the connection by the publisher (publishers). Only one type of port packet is accepted through the port connection. |
| | *fifo* | *fifo* connections: at the input port there is a circular fifo with a specific length. Port packets sent through the port connection by publishers, get stored there. Only one type of port packet is accepted through the port connection. |
| | *ufifo* | *unbounded fifo* connections: at the input port there is a fifo with a specific length. Port packets sent through the port connection by publishers, get stored there. When the fifo is full and port packets keep coming the fifo grows in order to store them. Only one type of port packet is accepted through the port connection. |
| *multipacket* | *multipacket* | *multipacket* connections: those connections accept a set of port packet types. There is a different buffer for each accepted port packet type, the last port packet of each type which is sent through the connection by publishers gets stored on them. |
| *lazymultipacket* | | *lazymultipacket* connections: those connections accept a set of port packet types. At the input port there is a different buffer for each accepted port packet type, the last port packet of each type which is sent through the connection by publishers gets stored on them. On the output port, port packets get stored in a queue of port packets to send, they are really sent to the other end when a *flush* operation is applied by the publisher on the output port. |

Table 1. Available port connections types: *active publisher/passive subscriber (AP/PS)* connections.

| Passive Publisher/Active Subscriber (PP/AS) | | |
|---|---|---|
| Output Port | Input Port | Port Connection Type |
| *poster* | *poster* | *poster* connections: there is a buffer at the output port where the last packet send by the publisher gets stored. There is another buffer at the input port which is "synchronized" with the output port buffer when the subscriber accesses its input port in order to get the last port packet sent through the connection. Therefore, the port packet gets copied to the input port only when a new port packet has been stored at the output port end. Only one type of port packet is accepted through the port connection. |

Table 2. Available port connections types: *passive publisher/ active subscriber (PP/AS)* connections.

packets and keeps doing something different, the subscriber gets packets at its own pace and not necessarily at the right moment they get to its input ports.

As to port packets, when defining an output or input port, we have to specify which port packet type or types (depending on the port type being defined), the port will accept. In general, port packet types are defined by the user, as we will see in section 3, we may also use port packets types provided by CoolBOT itself (the available ones are shown in Fig. 3), port packet types previously developed, or third party port packet types.

| Port packet type | Description |
|---|---|
| PacketUChar | Transports a C++ unsigned char. |
| PacketInt | Transports a C++ int. |
| PacketLong | Transports a C++ long. |
| PacketDouble | Transports a C++ double. |
| PacketTime | Transports a CoolBOT Time value (a time-stamp). |
| PacketCoordinates2D | Transports a CoolBOT Coordinates2D value (stores a 2D point). |
| PacketFrame2D | Transports a CoolBOT Frame2D value (stores a 2D frame). |
| PacketCoordinates3D | Transports a CoolBOT Coordinates3D value (stores a 3D point). |
| PacketFrame3D | Transports a CoolBOT Frame3D value (stores a 3D frame). |

Table 3. Available port packet types provided by CoolBOT itself.

## 2.2 CoolBOT components

CoolBOT components are *active objects* [Ellis & Gibbs (1989)], as [Brugali & Shakhimardanov (2010)] states: "a component is a computation unit that encapsulates data structures and operations to manipulate them". Moreover, in CoolBOT, components can be seen as "data-flow machines", since they process data when they dispose of it at their input ports. Otherwise, they stay idle, waiting for incoming port packets. On the other side, components send processed data in form of port packets through their output ports. All in all, the model of computation of CoolBOT systems follows the *Flow Based Programming* (FBP) paradigm according to [J. Paul Morrison (2010)], so systems can be built as networks of components interconnected by means of port connections. More formally, CoolBOT components are modeled as *port automata* [Steenstrup et al. (1983)][Stewart et al. (1997)]. Fig. 4 provides a view of the structure of a component in CoolBOT. There is a clear separation between its external interface and its internal structure. Externally, a component can only communicate with other components (and views and probes) through its input and output ports. Thence, a component's external interface comprises all its input and output ports, its types, and the port packets it accepts through them. As we can see on the figure there are two special ports in any component: the *control port* and the *monitoring port*, the rest of the ports are user defined. The *control port* allows to modify component's *controllable variables*. Through the *monitoring port* components publish their *observable variables*. Both ports allows an external supervisor (i.e. another component, a view or a probe) to observe and modify the execution and configuration of a given component.

Internally a component is organized as an automaton, as illustrated in Fig. 4. All components follow the same automaton structure. The part of this automaton which is common to all components is called the *default automaton*, and comprises several states, namely: *starting*, *ready*, *suspended*, *end*, and four states for component exception handling. This structure allows for an external supervisor to control the execution of any component in a system using an standard protocol, likewise in an operating system where threads and processes transit among different states during their lifetime. To complete the automaton of a component the user defines the *user automaton* which is specific for each component and implements its functionality. This is represented in Fig. 4 with a dotted circle as the meta-state *running*. Transitions among component's automaton states are triggered by incoming port packets through any of its input ports, and also by internal events (timers, empty transition, a control variable that has been modified by an external supervisor, entering or exiting a state, etc.). The user can associate C++ callbacks to transitions, much like the *codels* for $G^{en}$oM [Mallet et al. (2010)] modules.

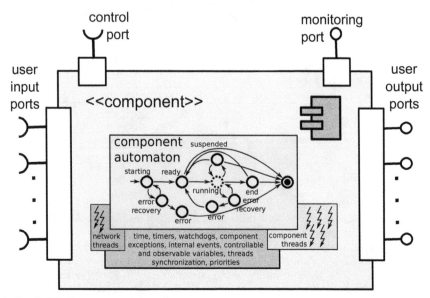

Fig. 4. CoolBOT. Component structure.

A key design principle for CoolBOT components is to take advantage of multithreaded and multicore capacities of mainstream operating systems, and the infrastructure they provide for multithreaded programming. Another key principle of design for components was to separate functional logic from thread synchronization logic. The user should be only worried about the functional logic, synchronization issues should be completely transparent to the developer. CoolBOT should be responsible for them behind the scenes. As active objects, CoolBOT components can organize its execution using multiple threads of execution as depicted on Fig. 4. Those threads are mapped on the underlying operating system (see Fig. 5). Thus, when developing a component the user assigns disjointly threads to automaton states, and to input ports and internal events provoking transitions. Those transitions, i.e. their associated callbacks, will be executed by the specific threads being assigned. The synchronization among them is guaranteed by the underlying framework infrastructure. All components are endowed at least with one thread of execution; the rest, if any, are user defined.

As depicted in Fig. 1, CoolBOT provides means for distributed computation. A given system can be mapped on a set formed by different machines sharing a computer network. Port connections among components, views and probes are transparently multiplexed using TCP/IP connections (see section 4). Furthermore, each machine can host one or several CoolBOT *integrations*. A CoolBOT *integration* is an application (a process) which integrates instances of CoolBOT components, views and probes. Integrations can be instantiated in any machine, and are user defined using a description language as we will see in next section.

## 3. CoolBOT development tools

CoolBOT provides several tools for helping developers. Fig. 6 shows the main ones, namely: `coolbot-ske` and `coolbot-c`. The former one, `coolbot-ske`, is used to create a directory structure for development of CoolBOT components, probes, port packets, views and integrations. It also generates CMake [Kitware, Inc. (2010)] template files for compiling them, description language template files for `coolbot-c`, and test programs for components. The

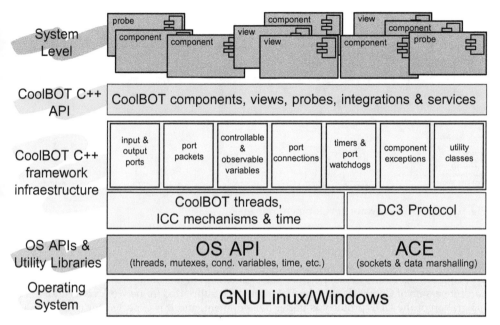

Fig. 5. Abstraction layers in CoolBOT.

latter tool, the CoolBOT compiler `coolbot-c`, generates C++ skeletons for components, port packets, views and integrations, and for each component it also generates its corresponding probe. All, the probe and the skeletons are C++ classes, and CoolBOT uses a description language as source code to generate those C++ classes. Except for probes, which are complete functional C++ classes, `coolbot-c` generates incomplete C++ classes which constitute the mentioned C++ skeletons. They are incomplete in the sense that they lack functionality, the user is responsible for completing them. Once completed, and using the CMake templates provided by `coolbot-ske`, they can be compiled. Components, probes, port packets, and views compile yielding dynamic libraries, integrations compile yielding executable programs. Moreover, the `coolbot-c` compiler preserves information when recompiling description files which have been modified, in such a way that, all C++ code introduced by the user into the skeletons is preserved.

## 4. Transparent distributed computation

Transparent distributed computation is the first development we have integrated on CoolBOT in order to improve its use and operation. The main idea was to make network communications as transparent as possible to developers (and components). We wanted CoolBOT to be responsible for them on behalf of components. Thus, at system level, to connect two component instances instantiated in different CoolBOT integrations should be as easy as connecting them when instantiated in the same integration. In particular, we follow three main principles related to transparent distributed computation facilities: transparent network inter component communications, network decoupling of component's functional logic, and incremental design.

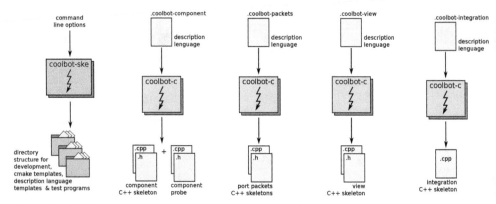

Fig. 6. CoolBOT's software development process.

### 4.1 Transparent network inter component communications

In order to make network communications transparent to components, we have developed a protocol termed *Distributed CoolBOT Component Communication Protocol (DC3P)* to multiplex port connections over TCP connections established among the components involved. In the current version of CoolBOT, only the TCP protocol is supported for network connections. The integration of the UDP protocol is under development, and it is expected for next CoolBOT version. DC3P has been implemented using the TCP/IP socket wrappers and the marshalling facilities provided by the ACE library [Douglas C. Schmidt (2010)], illustrated in Fig. 5. The protocol consists of the following packets:

- *Port Type Info* (request & response): For asking type information about input and output ports through network connections. This allows port connection compatibility verification when establishing a port connection through the network.

- *Connect* (request & response): For establishing a port connection over TCP/IP.

- *Disconnect* (request & response): To disconnect a previous established port connection.

- *Data Sending*: Once a port connection is established over TCP/IP, port packets are sent through it using this DC3P packet.

- *Remote Connect* (request & response): For establishing port connections between two remote component instances. Permits to connect component instances remotely.

- *Remote Disconnect* (request & response): To disconnect port connections previously established between two remote component instances.

- *Echo Request & Response*: Those DC3P packets are useful to verify that the other end in a network communication is active and responding.

All DC3P packets and port packets sent through port connections are marshalled and unmarshalled in order to be correctly sent through the network. We have used the facilities ACE provides for marshalling/demarshalling based on the OMG Common Data Representation (CDR) [Object Management Group (2002b)]. In general, port packets are user defined. In order to make their marshalling/demarshalling as easy and transparent as possible for developers, the description language accepted by the `coolbot-c` compiler includes sentences for describing port packets (as we can observe in Fig. 6), much like CORBA IDL [Object Management Group (2002a)]. The compiler generates a C++ skeleton class for each port packet where the code for marshalling/demarshalling is part of the skeleton's code generated automatically. In addition, we have endowed also CoolBOT with a rich set of C++ templates and classes to support marshalling and demarshalling of port packets (or any other arbitrary C++ class).

## 4.2 Network decoupling of component's functional logic
Another important aspect for network communication transparency is the decoupling of network communication logic from the functional logic of the component. Fig. 4 illustrates how this decoupling has been put into practice. Each component is endowed with a pair of network threads, and *output network thread*, and an *input network thread*, which are responsible for network communications using DC3P. CoolBOT guarantees transparently thread synchronization between them and the functional threads of the component. The network threads are mainly idle, waiting to have port packets to send through open network port connections, or to receive incoming port packets that should be redirected to the corresponding component's input ports. At instantiation time, it is possible to deactivate the network support for a component instance (and also for views and probes instances). In this manner, the component is not reachable from outside the integration where it has been instantiated, and evidently network threads and the resources they have associated are not allocated.

## 4.3 Incremental design
In future versions of CoolBOT, it is very possible that the set of DC3P protocol packets grow with new ones. In order to allow an easy integration of new DC3P packets in CoolBOT, we have applied the *composite* and *prototype* patterns [Gamma et al. (1995)] to their design. Those design patterns, jointly with the C++ templates and classes to support marshalling and demarshalling provide a systematic and easy manner of integrating new DC3P packets in future versions of the framework.

## 5. Deeper interface decoupling: Views and probes

Inspired by one of the "good practices" proposed by the authors of Carnegie Mellon's Navigation Toolkit CARMEN [Montemerlo et al. (2003)]: "one important design principle of CARMEN is the separation of robot control from graphical displays", we have introduced in CoolBOT the concept of *view* as an integrable, composite and reusable graphical interface available for CoolBOT system integrators and developers. Thus, CoolBOT *views* are graphical interfaces which, as software components, may be interconnected with any other component, view or probe in a CoolBOT system. In Fig. 7 is depicted the structure of a view in CoolBOT. As shown, CoolBOT views are also endowed with an external interface of input and output ports. Through this interface the view can communicate with other components, views and probes present in a given system. Identically to components, views are provided with the same network thread support which allows transparent and decoupled network communications through port connections. Internally, a view is a graphical interface, in fact, the current views already developed and operational which are available in CoolBOT have been implemented using the GTK graphical library [*The GTK+ Project* (2010)]. As shown in Fig. 6, C++ skeletons for views are generated using the `coolbot-c` compiler. The part which should be completed by the user is precisely the graphical implementation, which can be done using directly the GTK library or a GUI graphical programming software for designing window-based interfaces like Glade [*Glade - A User Interface Designer* (2010)].
As depicted in Fig. 7 a CoolBOT *probe* is provided with an external interface of input and output ports, and likewise component and views, as software components, this allows them to

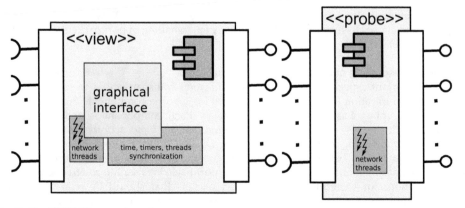

Fig. 7. CoolBOT. View and probe structures.

be interconnected with other components, views or probes. Equally they implement the same network decoupled support of threads for transparent network communications. In CoolBOT, *probes* are devised as interfaces for interoperability with non CoolBOT software, as illustrated graphically in Fig. 1. A complete functional C++ class implementing a probe is generated when a component is compiled by coolbot-c. The probe implements the complementary external interface of their corresponding component. Those probes generated automatically can be seen as automatic refactorings of external component interfaces in order to support interoperability of CoolBOT components with non CoolBOT software. As mentioned in [Makarenko et al. (2007)] this is an important factor in order to facilitate integration of different development robotic software approaches.

## 6. A real integration

In its current operating version, CoolBOT has been mainly used to control mobile robotic systems with the platforms we have available at our laboratory: Pioneer mobile robots models 3-DX, and P3-AT from Adept Mobile Robots [*Adept Mobile Robots* (2011)] (former Activ Media Robotics).

In this section we will show next a real robotic system using one of our Pioneer 3-DX mobile robots, in order to give a glimpse of a real system controlled using CoolBOT. The system is illustrated in Fig. 8. The example shows a secure navigation system for an indoor mobile robot. This is a real application we have usually in operation on the robots we have at the laboratory. The system implements a secure navigation system based on the ND+ algorithm for obstacle avoidance [Minguez et al. (2004)]. It has been implemented attending to [Montesano et al. (2006)]. In the figure, input ports, output ports, and port connections, have been reduced for the sake of clarification. Some of them represent several connections and ports in the real system.

The system is organized using two CoolBOT integrations, one only formed by CoolBOT component instances, and the other one containing CoolBOT view instances. The former one is really the integration which implements the navigation system. As we can observe, it consists of five component instances, namely: PlayerRobot (this is a wrapper component for hardware abstraction using the Player/Stage project framework [Vaughan et al. (2003)]), MbICP (this is a component which implements the MbICP scan matching algorithm based on laser range sensor data [Minguez et al. (2006)]), GridMap (this component maintains a grid map of the surroundings of the robot built using robot range laser scans, it also generates

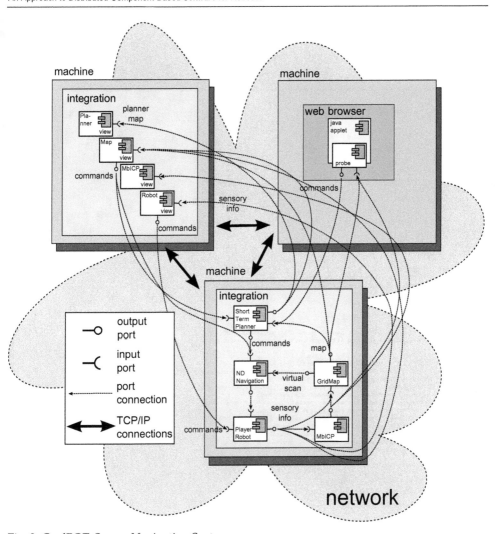

Fig. 8. CoolBOT. Secure Navigation System.

periodically a 360° virtual scan for the ND+ algorithm), NDNavigation (implements the ND+ algorithm) and ShortTermPlanner (a planner which uses the grid map for planning paths in the robot surroundings using a modification of the numerical navigation function NF2 found in [Jean-Claude Latombe (1991)]). On other machine another integration is shown hosting four view instances through which we can control and monitor the system remotely. In addition, in another machine, there is a web browser hosting a Java applet using a CoolBOT *probe* to connect to some of the components of the system.

In order to clarify how the integration of Fig. 8 has been built, and also to clarify the process of development of each of its components, in next paragraphs we will have a look at the description files used to generate some of them, including the whole integration shown in the figure. Thus, in Fig. 9 we can see the description file accepted by coolbot-c of one of the

```
/*
 * File: player-robot.coolbot
 * Description: description file for PlayerRobot component
 * Date: 02 June 2010
 * Generated by coolbot-ske
 */

component PlayerRobot
{
  header
  {
    author "Antonio Carlos Domínguez Brito <adominguez@iusiani.ulpgc.es>";
    description "PlayerRobot component";
    institution "IUSIANI - Universidad de Las Palmas de Gran Canaria";
    version "0.1"
  };

  constants
  {
    LASER_MAX_RANGE="LaserPacket::LASER_MAX_RANGE";
    SONAR_MAX_RANGE=5000; // millimeters

    private FIFO_LENGTH=5;
    private ROBOT_DATA_INCOMING_FREQUENCY= 10; // Hz
    private LASER_MIN_ANGLE= -90; // degrees

    ...
  };

  // input ports
  input port Commands type fifo port packet CommandPacket length FIFO_LENGTH;
  input port NavigationCommands type fifo port packet NDNavigation::CommandPacket length FIFO_LENGTH;

  //output ports
  output port RobotConfig type poster port packet ConfigPacket;
  output port Odometry type generic port packet OdometryPacket network buffer FIFO_LENGTH;
  output port OdometryReset type generic port packet PacketTime;
  output port BumpersGeometry type poster port packet "BumperGeometryPacket";
  output port Bumpers type generic port packet BumperPacket;
  output port SonarGeometry type poster port packet "SonarGeometryPacket";
  output port SonarScan type generic port packet "SonarPacket";
  output port LaserGeometry type poster port packet PacketFrame3D;
  output port LaserScan type generic port packet LaserPacket;
  output port Power type generic port packet PacketDouble;
  output port CameraImage type poster port packet CameraImagePacket;
  output port PTZJoints type generic port packet "PacketPTZJoints";

  exception RobotConnection
  {
    description "Robot connection failed.";
  };

  exception NoPositionProxy
  {
    description "Position proxy not available in this robot.";
  };

  exception InternalPlayerException
  {
    description "A Player library exception has been thrown.";
  };

  entry state Main
  {
    transition on Commands,NavigationCommands,Timer;
  };
};
```

Fig. 9. `player-robot.coolbot-component`: `PlayerRobot`'s description file.

components of Fig. 8, file `player-robot.coolbot-component`, corresponding to component `PlayerRobot`. As to views, in Fig. 10, we can see the description file for one of the view instances of Fig. 8, concretely for the `Map` view in the figure, which is an instance of view `GridGtk`. As we can observe, the description file specifies mainly the view's external interface formed by input and output ports.

```
/*
 * File: grid-gtk.coolbot-view
 * Description: description file for GridGtk view
 * Date: 29 April 2011
 * Generated by coolbot-ske
 */

view GridGtk
{
  header
  {
    author "Antonio Carlos Dominguez-Brito <adominguez@iusiani.ulpgc.es>";
    description "GridGtk View";
    institution "IUSIANI - ULPGC (Spain)";
    version "0.1"
  };

  constants
  {
    private DEFAULT_REFRESHING_PERIOD=500; // milliseconds
    ...
  };

  // input ports
  input port ROBOT_CONFIG type poster port packet PlayerRobot::ConfigPacket;
  input port GRID_MAP type poster port packet GridMap::GridMapPacket;
  input port PLANNER_PATH type last port packet ShortTermPlanner::PlannerPathPacket;

  // output ports
  output port PLANNER_COMMANDS type generic port packet ShortTermPlanner::CommandPacket;
  output port ND_COMMANDS type generic port packet NDNavigation::CommandPacket;
};
```

Fig. 10. `grid-gtk.coolbot-view`: `GridGtk`'s description file.

In Fig. 11 it is shown a snapshot of the view in runtime. Once developed CoolBOT views are graphical components we can integrate in a window-based application like the one shown in the figure. In particular, in the application we can see, views are plugged in as "pages" of the GTK widget `notepad` (a container of "pages" with "tabs") we can observe in the figure. In fact, the GUI application shown integrates the four view instances of Fig. 8, whose C++ skeleton has been generated using also `coolbot-c` from a description file (the `.coolbot-integration` file in Fig. 6). In Fig. 12 we can see part of this file. Notice that `coolbot-c` generates C++ skeletons for integrations where the static instantiation and interconnection among components and views are generated automatically. If we want to build a dynamic integration, in terms of dynamic instantiation of components and views, and also in terms of dynamic interconnections among them establishing port connections, we must complete the dynamic part of the skeleton generated by `coolbot-c` using the C++ runtime services provided by the framework (Fig. 5).

With respect to CoolBOT probes, as of now we have used them to interoperate with Java applets inserted in a web browser, as shown in Fig. 8. More specifically we have used SWIG [*SWIG* (2011)] in order to have access to the probe C++ classes in Java with the aim of implementing several Java GUI interfaces in Java equivalent to some of the CoolBOT views we have already developed. In Fig. 13 we can see a snapshot of the Java equivalent of a view to represent the range sensor information of a mobile robot.

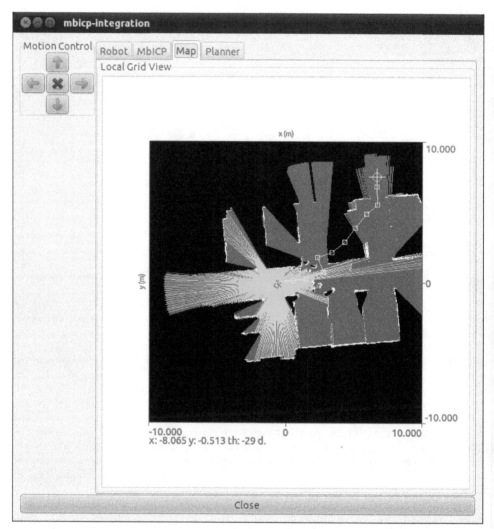

Fig. 11. View GridGtk's snapshot.

```
/*
 * File: mbicp-integration.coolbot-integration
 * Description: description file for mbicp-integration integration.
 * Date: 29 April 2011
 * Generated by coolbot-ske
 */

integration mbicp_integration
{
  header
  {
    author "Antonio Carlos Domínguez-Brito <adominguez@iusiani.ulpgc.es>";
    description "MbICP's views integration";
    institution "IUSIANI - ULPGC (Spain)";
    version "0.1"
  };

  machine addresses
  {
    local my_machine: "127.0.0.1";
    the_other_machine: "...";
  };

  listening ports // TCP/IP ports
  {
    ROBOT_PORT: 1950;
    MBICP_PORT: 1965;
    NAVIGATION_MAP_PORT: 1970;
    ND_PORT: 1980;
    NAVIGATION_PLANNER_PORT: 1990;

    ROBOT_VIEW_PORT: 1955;
    MBICP_VIEW_PORT: 1985;
    NAVIGATION_MAP_VIEW_PORT: 1975;
    NAVIGATION_PLANNER_VIEW_PORT: 1995;
  };

  local instances
  {
    view robotView:PlayerRobotGtk listening on ROBOT_VIEW_PORT with description "Robot";
    view mbicpView:MbICPGtk listening on MBICP_VIEW_PORT with description "MbICP";
    view mapView:GridGtk listening on NAVIGATION_MAP_VIEW_PORT with description "Map";
    view navigationPlannerView:PlannerGtk listening on NAVIGATION_PLANNER_VIEW_PORT with description "Planner";
  };

  remote instances on the_other_machine;
  {
    component robot:PlayerRobot listening on ROBOT_VIEW_PORT;
    component mbicpInstance:MbICPCorrector listening on MBICP_PORT;
    component navigationMap:GridMap listening on NAVIGATION_MAP_PORT;
    component nd:NDNavigation listening on ND_PORT;
    component navigationPlanner:ShortTermPlanner listening on NAVIGATION_PLANNER_VIEW_PORT;
  };

  port connections // static connections
  {
    connect robot:ODOMETRY to robotView:ODOMETRY;
    connect robot:LASERGEOMETRY to robotView:LASER_GEOMETRY;
    connect robot:LASERSCAN to robotView:LASER_SCAN;
    connect robot:POWER to robotView:POWER;
    connect robot:SONARGEOMETRY to robotView:SONAR_GEOMETRY;
    connect robot:SONARSCAN to robotView:SONAR_SCAN;

    connect robot:ODOMETRY to mbicpInstance:ODOMETRY;
    connect robot:LASERGEOMETRY to mbicpInstance:LASER_GEOMETRY;
    connect robot:LASERSCAN to mbicpInstance:LASER_SCAN;

    connect robotView:COMMANDS to robot:COMMANDS;

    ...
  };
};
```

Fig. 12. The integration containing Robot. MbICP, Map and Planner views of Fig. 8

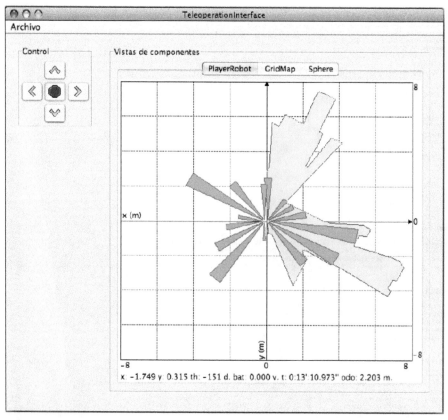

Fig. 13. Java view implemented using a probe to access range sensor information for a mobile robot. A snapshot.

# 7. Conclusions

In this document we have presented the last developments which have been integrated in the last operating version of CoolBOT. The developments have been aimed mainly to two questions: transparent distributed computation, and "deeper" interface decoupling. It is our opinion that the use and operation of CoolBOT has improved considerably. CoolBOT is an open source initiative supported by our laboratory which is freely available via **www.coolbotproject.org**, including the secure navigation system depicted in Fig. 8.

# 8. References

*Adept Mobile Robots* (2011). http://www.mobilerobots.com/.

Ando, N., Suehiro, T. & Kotoku, T. (2008). A Software Platform for Component Based RT-System Development: OpenRTM-Aist, *in* S. Carpin, I. Noda, E. Pagello, M. Reggiani & O. von Stryk (eds), *Simulation, Modeling, and Programming for Autonomous Robots*, Vol. 5325 of *Lecture Notes in Computer Science*, Springer Berlin / Heidelberg, pp. 87–98.

Antonio C. Domínguez-Brito, Daniel Hernández-Sosa, José Isern-González & Jorge Cabrera-Gámez (2007). *Software Engineering for Experimental Robotics*, Vol. 30 of *Springer Tracts in Advanced Robotics Series*, Springer, chapter CoolBOT: a Component Model and Software Infrastructure for Robotics, pp. 143–168.

Basu, A., Bozga, M. & Sifakis, J. (2006). Modeling heterogeneous real-time components in BIP, In Fourth IEEE International Conference on Software Engineering and Formal Methods, pages 3-12, Pune (India).

Bensalem, S., Gallien, M., Ingrand, F., Kahloul, I. & Thanh-Hung, N. (2009). Designing autonomous robots, *IEEE Robotics and Automation Magazine* 16(1): 67–77.

Brooks, A., Kaupp, T., Makarenko, A., S.Williams & Oreback, A. (2005). Towards component-based robotics, *In IEEE International Conference on Intelligent Robots and Systems*, Tsukuba, Japan, pp. 163–168.

Brugali, D. (ed.) (2007). *Software Engineering for Experimental Robotics*, Springer Tracts in Advanced Robotics, Springer.

Brugali, D. & Scandurra, P. (2009). Component-based robotic engineering (part i) [tutorial], *Robotics Automation Magazine, IEEE* 16(4): 84 –96.

Brugali, D. & Shakhimardanov, A. (2010). Component-based robotic engineering (part ii), *Robotics Automation Magazine, IEEE* 17(1): 100 –112.

Domínguez-Brito, A. C., Hernández-Sosa, D., Isern-González, J. & Cabrera-Gámez, J. (2004). Integrating Robotics Software, IEEE International Conference on Robotics and Automation, New Orleans, USA.

Douglas C. Schmidt (2010). The Adaptive Communication Environment (ACE), www.cs.wustl.edu/~schmidt/ACE.html.

Ellis, C. & Gibbs, S. (1989). *Object-Oriented Concepts, Databases, and Applications*, ACM Press, Addison-Wesley, chapter Active Objects: Realities and Possibilities.

Gamma, E., Helm, R., Johnson, R. & Vlissides, J. (1995). *Design Patterns: Elements of Reusable Object-Oriented Software*, Addison-Wesley Professional Computing Series, Addison-Wesley.

George T. Heineman & William T. Councill (2001). *Component-Based Software Engineering*, Addison-Wesley.

*Glade - A User Interface Designer* (2010). glade.gnome.org.

J. Paul Morrison (2010). *Flow-Based Programming, 2nd Edition: A New Approach to Application Development*, CreateSpace.

Jean-Claude Latombe (1991).   *Robot-motion planning*, The Kluwer International Series in Engineering and Computer Science, Kluwer Academic.

Kitware, Inc. (2010). The CMake Open Source Build System, www.cmake.org.

Makarenko, A., Brooks, A. & Kaupp, T. (2007). On the benefits of making robotic software frameworks thin, IEEE/RSJ Int. Conf. on Intelligent Robots and Systems (IROS'07), San Diego CA, USA.

Mallet, A., Pasteur, C., Herrb, M., Lemaignan, S. & Ingrand, F. (2010). GenoM3: Building middleware-independent robotic components, IEEE International Conference on Robotics and Automation.

Minguez, J., Montesano, L. & Lamiraux, F. (2006).   Metric-based iterative closest point scan matching for sensor displacement estimation, *Robotics, IEEE Transactions on* 22(5): 1047 –1054.

Minguez, J., Osuna, J. & Montano, L. (2004). A "Divide and Conquer" Strategy based on Situations to achieve Reactive Collision Avoidance in Troublesome Scenarios, IEEE International Conference on Robotics and Automation, New Orleans, USA.

Montemerlo, M., Roy, N. & Thrun, S. (2003). Perspectives on standardization in mobile robot programming: the carnegie mellon navigation (carmen) toolkit, Vol. 3, pp. 2436 – 2441 vol.3.

Montesano, L., Minguez, J. & Montano, L. (2006).   Lessons Learned in Integration for Sensor-Based Robot Navigation Systems, *International Journal of Advanced Robotic Systems* 3(1): 85–91.

Object      Management      Group      (2002a).            OMG      IDL:      Details, (http://www.omg.org/gettingstarted/omg_ idl.htm).

Object Management Group (2002b). The Common Object Request Broker: Architecture and Specification, Ch. 15, Sec. 1-3. (http://www.omg.org/cgi-bin/doc?formal/02-06-01).

*ROS: Robot Operating System* (2011). http://www.ros.org.

Schlegel, C., Haßler, T., Lotz, A. & Steck, A. (2009).   Robotic Software Systems: From Code-Driven to Model-Driven Designs, In Proc. 14th Int. Conf. on Advanced Robotics (ICAR), Munich.

Steenstrup, M., Arbib, M. A. & Manes, E. G. (1983).   Port automata and the algebra of concurrent processes, *Journal of Computer and System Sciences* 27: 29–50.

Stewart, D. B., Volpe, R. A. & Khosla, P. (1997).   Design of Dynamically Reconfigurable Real-Time Software Using Port-Based Objects, *IEEE Transactions on Software Engineering* 23(12): 759–776.

*SWIG* (2011). http://www.swig.org/.

*The GTK+ Project* (2010). www.gtk.org.

*The Orocos Project* (2011). http://www.orocos.org.

Vaughan, R. T., Gerkey, B. & Howard, A. (2003).   On Device Abstractions For Portable, Reusable Robot Code, *IEEE/RSJ International Conference on Intelligent Robot Systems (IROS 2003), Las Vegas, USA, October 2003*, pp. 2121–2427.

# 9

# Methodology for System Adaptation Based on Characteristic Patterns

Eva Volná[1], Michal Janošek[1], Václav Kocian[1],
Martin Kotyrba[1] and Zuzana Oplatková[2]
*[1]University of Ostrava*
*[2]Tomas Bata University in Zlín*
*Czech Republic*

## 1. Introduction

This paper describes the methodology for system description and application so that the system can be managed using real time system adaptation. The term system here can represent any structure regardless its size or complexity (industrial robots, mobile robot navigation, stock market, systems of production, control systems, etc.). The methodology describes the whole development process from system requirements to software tool that will be able to execute a specific system adaptation.

In this work, we propose approaches relying on machine learning methods (Bishop, 2006), which would enable to characterize key patterns and detect them in real time and in their altered form as well. Then, based on the pattern recognized, it is possible to apply a suitable intervention to system inputs so that the system responds in the desired way. Our aim is to develop and apply a hybrid approach based on machine learning methods, particularly based on soft-computing methods to identify patterns successfully and for the subsequent adaptation of the system. The main goal of the paper is to recognize important pattern and adapt the system's behaviour based on the pattern desired way.

The paper is arranged as follows: Section 1 introduces the critical topic of the article. Section 2 details the feature extraction process in order to optimize the patterns used as inputs into experiments. The pattern recognition algorithms using machine learning methods are discussed in section 3. Section 4 describes the used data-sets and covers the experimental results and a conclusion is given in section 5. We focus on reliability of recognition made by the described algorithms with optimized patterns based on the reduction of the calculation costs. All results are compared mutually.

### 1.1 The methodology for system description

Gershenson (Gershenson, 2007) proposed a methodology called *The General Methodology* for system description necessary to manage a system. It presents a conceptual framework for describing systems as self-organizing and consists of five steps: representation, modelling, simulation, application and evaluation. Our goal is to use and adapt this methodology for our specific needs. Basically we would like to describe a methodology that the designer should be able to use to describe his system, find key patterns in its behaviour based on the observation and prepare suitable response to these patterns that emerge from time to time and adapt to any deviation in the system's behaviour.

As we are using Gershenson's methodology we are not going to describe it in detail because detailed info can be found in his book (Gershenson, 2007). Let's mention crucial parts of his methodology that is important to our work. The methodology is useful for designing and controlling complex systems. Basically a complex system consists of two or more interconnected components and these components react together and it is very complicated to separate them. So the system's behaviour is impossible to deduce from the behaviour of its individual components. This deduction becomes more complicated how more components $\#\overline{E}$ and more interactions $\#\overline{I}$ the system has ($C_{sys}$ corresponds with system complexity; $C_e$ corresponds with element complexity; $C_i$ corresponds with interaction complexity).

$$
C_{sys} \sim \left\{ 
\begin{array}{l}
\overline{E} \\[4pt]
\overline{I} \\[4pt]
\displaystyle\sum_{j=0}^{\#\overline{E}} C_{e_j} \\[12pt]
\displaystyle\sum_{k=0}^{\#\overline{I}} C_{i_k}
\end{array}
\right. \tag{1}
$$

Imagine a manufacturing factory. We can describe the manufacturing factory as a complex system. Now it is important to realize that we can have several levels of abstraction starting from the manufacturing line to the whole factory complex. The manufacturing line can consist of many components. There can be robots, which perform the main job. Conveyor belts, roller beds, jigs, hangers and other equipment responsible for the product or material transport and other equipments. All the interactions are some way related to the material or product. Although it is our best interest to run all the processes smoothly there will be always some incidents we cannot predict exactly. The supply of the material can be interrupted or delayed, any equipment can have a multifunction and it is hard to predict when and how long will it takes. Because there are interactions among many of these components we can call manufacturing factory a complex system.

If we want to characterize a system we should create its model. Gershenson (Gershenson, 2002) proposes two types of models, absolute and relative. The absolute model (abs-model) refers to what the thing actually is, independently of the observer. The relative model (rel-model) refers to the properties of the thing as distinguished by an observer within a context. We can say that the rel-model is a model, while the abs-model is modelled. Since we are all limited observers, it becomes clear that we can speak about reality only with rel-beings/models (Gershenson, 2007).

So how we can model a complex system? Any complex system can be modelled using multi-agent system (MAS) where each system's component is represented by an agent and any interactions among system's components are represented as interactions among agents. If we take into consideration *The General Methodology* thus any system can be modelled as group of agents trying to satisfy their goals. There is a question. Can we describe a systems modelling as a group of agents as self-organizing? We think that we can say *Yes*. As the agents in the MAS try to satisfy their goals, same as components in self-organizing systems interact with each other to achieve desired state or behaviour. If we determine the state as a self-organizing state, we can call that system self-organizing and define our complex self-organizing system. In our example with manufacturing line our self-organizing state will be a state where the production runs smoothly without any production delays. But how can we achieve that?

Still using Gerhenson's General Methodology we can label fulfilling agent's goal as its satisfaction $\sigma \in [0,1]$. Then the system's satisfaction $\sigma_{sys}$ (2) can be represented as function $f : \mathbf{R} \to [0,1]$ and it is a satisfaction of its individual components.

$$\sigma_{sys} = f(\sigma_1, \sigma_2, ..., \sigma_n, w_0, w_1, w_2, ..., w_n) \tag{2}$$

$w_0$ represents bias and other weights $w_i$ represents an importance given to each $\sigma_i$.

Components, which decrease $\sigma_{sys}$ and increase their $\sigma_i$ shouldn't be considered as a part of the system. Of course it is hard to say if for higher system's satisfaction it is sufficient to increase satisfaction of each individual component because some components can use others fulfilling their goals. For maximization of $\sigma_{sys}$ we should minimize the friction among components and increase their synergy. A mediator arbitrates among elements of a system, to minimize conflict, interferences and frictions; and to maximize cooperation and synergy. So we have two types of agents in the MAS. Regular agents fulfil their goals and mediator agents streamline their behaviour. Using that simple agent's division we can build quite adaptive system.

## 1.2 Patterns as a system's behaviour description

Every system has its unique characteristics that can be described as patterns. Using patterns we would like to characterize particular system and its key characteristics. Generally a system can sense a lot of data using its sensors. If we put the sensor's data into some form, a set or a graph then a lot of patterns can be recognized and further processed. When every system's component has some sensor then the system can produce some patterns in its behaviour. Some sensor reads data about its environment so we can find some patterns of the environment, where the system is located. If we combine several sensors data, we would be able to recognise some patterns in the whole system's behaviour. It is important to realize that everything, which we observe is relative from our point of view. When we search for the pattern, we want to choose such pattern, which represents the system reliably and define its important properties. Every pattern, which we find, is always misrepresented with our point of view.

We can imagine a pattern as some object with same or similar properties. There are many ways how to recognize and sort them. When we perform pattern recognition, we assign a pre-defined output value to an input value. For some purpose, we can use a particular pattern recognition algorithm, which is introduced in (Ciskowski & Zaton, 2010). In this case we try to assign each input value to the one of the output sets of values. Some input value can be any data regardless its origin as a text, audio, image or any other data. When patterns repeat in the same or altered forms then can be classified into predefined classes of patterns. Since we are working on computers, the input data and all patterns can be represented in a binary form without the loss of generality. Such approach can work nearly with any system, which we would like to describe. But that is a very wide frame content.

Although theory of regulation and control (Armstrong & Porter, 2006) is mainly focused on methods of automatic control, it also includes methods for adaptive and fuzzy controls. In general, through the control or regulation we guide the system's behaviour in the desired direction. For our purposes, it suffices to regulate the system behaviour based on the predefined target and compensate any deviation in desired direction. So we search for key patterns in system's behaviour a try to adapt to any changes. However, in order to react quickly and appropriately, it is good to have at least an expectation of what may happen and which reaction would be appropriate, i.e. what to anticipate. Expectations are subjective probabilities that we learn from experience: the more often pattern B appears after pattern A, or the more successful action B is in solving problem A, the stronger the association A → B becomes. The next time we encounter A (or a pattern similar to A), we will be prepared, and more likely to react adequately. The simple ordering of options according to the probability that they would be relevant immensely decreases the complexity of decision-making (Heylighen, 1994).

Agents are appropriate for defining, creating, maintaining, and operating the software of distributed systems in a flexible manner, independent of service location and technology. Systems of agents are complex in part because both the structural form and the behaviour patterns of the system change over time, with changing circumstances. By structural form, we mean the set of active agents and inter-agent relationships at a particular time. This form changes over time as a result of inter-agent negotiations that determine how to deal with new circumstances or events. We call such changing structural form morphing, by analogy with morphing in computer animation. By behaviour patterns, we mean the collaborative behaviour of a set of active agents in achieving some overall purpose. In this sense, behaviour patterns are properties of the whole system, above the level of the internal agent detail or of pair wise, inter-agent interactions. Descriptions of whole system behaviour patterns need to be above this level of detail to avoid becoming lost in the detail, because agents are, in general, large grained system components with lots of internal detail, and because agents may engage in detailed sequences of interactions that easily obscure the big picture. In agent systems, behaviour patterns and morphing are inseparable, because they both occur on the same time scale, as part of normal operation. Use case maps (UCMs) (Burth & Hubbard, 1997) are descriptions of large grained behaviour patterns in systems of collaborating large grained components.

### 1.3 System adaptation vs. prediction

Let's say we have built pattern recognition system and it is working properly to meet our requirements. We are able to recognize certain patterns reliably. What can we do next? Basically, we can predict systems behaviour or we can adapt to any change that emerge.

It is possible to try to predict what will happen, but more or less it is a lottery. We will never be able to predict such systems' behaviour completely. This doesn't mean it is not possible to build a system based on prediction (Gershenson, 2007). But there is another approach that tries to adapt to any change by reflecting current situation. To adapt on any change (expected or unexpected) it should be sufficient to compensate any deviation from desired course. In case that response to a deviation comes quickly enough that way of regulation can be very effective. It does not matter how complicated system is (how many factors and interactions has) in case we have efficient means of control (Armstrong & Porter, 2006). To respond quickly and flexible it is desirable to have some expectation what can happen and what kind of response will be appropriate. We can learn such expectation through experiences.

## 2. Feature extraction process in order to optimize the patterns

Identification problems involving time-series data (or waveforms) constitute a subset of pattern recognition applications that is of particular interest because of the large number of domains that involve such data. The recognition of structural shapes plays a central role in distinguishing particular system behaviour. Sometimes just one structural form (a bump, an abrupt peak or a sinusoidal component), is enough to identify a specific phenomenon. There is not a general rule to describe the structure – or structure combinations – of various phenomena, so specific knowledge about their characteristics has to be taken into account. In other words, signal structural shape may be not enough for a complete description of system properties. Therefore, domain knowledge has to be added to the structural information.

However, the goal of our approach is not knowledge extraction but to provide users with an easy tool to perform a first data screening. In this sense, the interest is focused on searching for specific patterns within waveforms (Dormido-Canto et al., 2006). The algorithms used in pattern recognition systems are commonly divided into two tasks, as shown in Fig. 1. The description task transforms data collected from the environment into features (primitives).

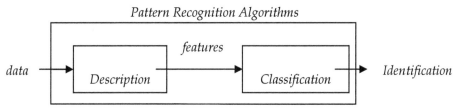

Fig. 1. Tasks in the pattern recognition systems

The classification task arrives at an identification of patterns based on the features provided by the description task. There is no general solution for extracting structural features from data. The selection of primitives by which the patterns of interest are going to be described depends upon the type of data and the associated application. The features are generally designed making use of the experience and intuition of the designer.

The input data can be presented to the system in various forms. In principle we can distinguish two basic possibilities:

• The numeric representation of monitored parameters
• Image data - using the methods of machine vision

Figures 2 and 3 show an image and a numerical expression of one particular section of OHLC data. The image expression contains only information from the third to the sixth column of the table (Fig.3). In spite of the fact, the pattern size (number of pixels) equals to 7440. In contrast to it, a table expression with 15 rows and 7 columns of 16-bit numbers takes only.

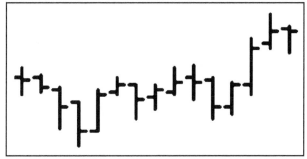

Fig. 2. Visual representations of pattern

| 2010.06.02 | 08:15 | 1.22220 | 1.22260 | 1.22140 | 1.22210 | 107 |
| 2010.06.02 | 08:20 | 1.22220 | 1.22230 | 1.22150 | 1.22170 | 76 |
| 2010.06.02 | 08:25 | 1.22160 | 1.22170 | 1.21990 | 1.22090 | 71 |
| 2010.06.02 | 08:30 | 1.22110 | 1.22110 | 1.21910 | 1.21970 | 85 |
| 2010.06.02 | 08:35 | 1.21980 | 1.22160 | 1.21970 | 1.22140 | 78 |
| 2010.06.02 | 08:40 | 1.22150 | 1.22220 | 1.22140 | 1.22190 | 61 |
| 2010.06.02 | 08:45 | 1.22180 | 1.22190 | 1.22030 | 1.22120 | 84 |
| 2010.06.02 | 08:50 | 1.22130 | 1.22180 | 1.22070 | 1.22160 | 71 |
| 2010.06.02 | 08:55 | 1.22150 | 1.22260 | 1.22140 | 1.22200 | 61 |
| 2010.06.02 | 09:00 | 1.22220 | 1.22270 | 1.22120 | 1.22200 | 91 |
| 2010.06.02 | 09:05 | 1.22210 | 1.22220 | 1.22030 | 1.22090 | 53 |
| 2010.06.02 | 09:10 | 1.22080 | 1.22200 | 1.22050 | 1.22190 | 87 |
| 2010.06.02 | 09:15 | 1.22180 | 1.22390 | 1.22140 | 1.22350 | 90 |
| 2010.06.02 | 09:20 | 1.22370 | 1.22500 | 1.22350 | 1.22430 | 87 |
| 2010.06.02 | 09:25 | 1.22440 | 1.22450 | 1.22330 | 1.22430 | 77 |

Fig. 3. Tabular expression of pattern

The image data better correspond to an intuitive human idea of patterns recognitions, which is their main advantage. We also have to remember that even table data must be transferred into binary (image) form before their processing.

Image data are always two-dimensional. Generally, tabular patterns can have more dimensions. Graphical representation of OHLC data (Lai, 2005) in Fig.2 is a good example of the expression of multidimensional data projection to two-dimensional space. Fig.2 shows a visual representation of 4-dimensional vector in time, which corresponds to 5 - dimensions. In this article, we consider experiments only over two-dimensional data (time series). Extending the principles of multidimensional vectors (random processes) will be the subject of our future projects.

The intuitive concept of "pattern" corresponds to the two-dimensional shapes. This way allows showing a progress of a scalar variable. In the case that a system has more than one parameter, the graphic representation is not trivial anymore.

## 3. Pattern recognition algorithms

Classification is one of the most frequently encountered decision making tasks of human activity. A classification problem occurs when an object needs to be assigned into a predefined group or class based on a number of observed attributes related to that object. Pattern recognition is concerned with making decisions from complex patterns of information. The goal has always been to tackle those tasks presently undertaken by humans, for instance to recognize faces, buy or sell stocks or to decide on the next move in a chess game. Rather simpler tasks have been considered by us. We have defined a set of classes, which we plan to assign patterns to, and the task is to classify a future pattern as one of these classes. Such tasks are called classification or supervised pattern recognition. Clearly someone had to determine the classes in the first phase. Seeking the groupings of patterns is called cluster analysis or unsupervised pattern recognition. Patterns are made up of features, which are measurements used as inputs to the classification system. In case that patterns are images, the major part of the design of a pattern recognition system is to select suitable features; choosing the right features can be even more important than what is done with them subsequently.

## 3.1 Artificial neural networks

Neural networks that allow so-called supervised learning process (i.e. approach, in which the neural network is familiar with prototype of patterns) use to be regarded as the best choice for pattern recognition tasks. After adaptation, it is expected that the network is able to recognise learned (known) or similar patterns in input vectors. Generally, it is true - the more training patterns (prototypes), the better network ability to solve the problem. On the other hand, too many training patterns could lead to exceeding a memory capacity of the network. We used typical representative of neural networks, namely:

- Hebb network
- Backpropagation network

Our aim was to test two networks with extreme qualities. In other words, we chose such neural networks, which promised the greatest possible differences among achieved results.

### 3.1.1 Hebb network

Hebb network is the simplest and also the "cheapest" neural network, which adaptation runs in one cycle. Both adaptive and inactive modes work with integer numbers. These properties allow very easy training set modification namely in applications that work with very large input vectors (e.g. image data).
Hebbian learning in its simplest form (Fausett, 1994) is given by the weights update rule (3)

$$\Delta w_{ij} = \eta \, a_i \, a_j \tag{3}$$

where $w_{ij}$ is the change in the strength of the connection from unit $j$ to unit $i$, $a_i$ and $a_j$ are the activations of units $i$ and $j$ respectively, and $\eta$ is a learning rate. When training a network to classify patterns with this rule, it is necessary to have some method of forcing a unit to respond strongly to a particular pattern. Consider a set of data divided into classes $C_1, C_2,...,C_m$. Each data point $x$ is represented by the vector of inputs $(x_1, x_2, ..., x_n)$. A possible network for learning is given in Figure 4. All units are linear. During training the class inputs $c_1, c_2, ...,c_m$ for a point $x$ are set as follows (4):

$$c_i = 1 \quad x \in C_i$$
$$c_i = 0 \quad x \notin C_i \tag{4}$$

Each of the class inputs is connected to just one corresponding output unit, i.e. $c_i$ connects to $o_i$ only for $i$ = 1, 2, ...,$m$. There is full interconnection from the data inputs $x_1, x_2, ..., x_n$ to each of these outputs.

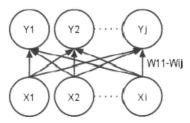

Fig. 4. Hebb network. Weights of connections w11-wij are modified in accordance with the Hebbian learning rule

### 3.1.2 Backpropagation network

Back propagation network is one of the most complex neural networks for supervised learning. Its ability to learning and recognition are much higher than Hebb network, but its disadvantage is relatively lengthy processes of adaptation, which may in some cases (complex input vectors) significantly prolong the network adaptation to new training sets. Backpropagaton network is a multilayer feedforward neural network. See Fig. 5, usually a fully connected variant is used, so that each neuron from the *n-th* layer is connected to all neurons in the *(n+1)-th* layer, but it is not necessary and in general some connections may be missing – see dashed lines, however, there are no connections between neurons of the same layer. A subset of input units has no input connections from other units; their states are fixed by the problem. Another subset of units is designated as output units; their states are considered the result of the computation. Units that are neither input nor output are known as hidden units.

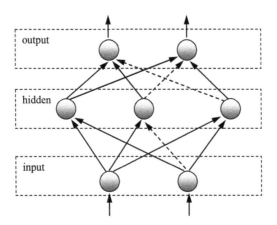

Fig. 5. A general three-layer neural network

Backpropagation algorithm belongs to a group called "gradient descent methods". An intuitive definition is that such an algorithm searches for the global minimum of the weight landscape by descending downhill in the most precipitous direction. The initial position is set at random selecting the weights of the network from some range (typically from -1 to 1 or from 0 to 1). Considering the different points, it is clear, that backpropagation using a fully connected neural network is not a deterministic algorithm. The basic backpropagation algorithm can be summed up in the following equation (the *delta rule*) for the change to the weight $w_{ji}$ from node $i$ to node $j$ (5):

$$
\begin{array}{ccccc}
\text{weight} & \text{learning} & \text{local} & \text{input signal} & \\
\text{change} & \text{rate} & \text{gradient} & \text{to node } j & \quad (5)\\
\Delta w_{ji} & = \quad \eta & \times \quad \delta_j & \times \quad y_i &
\end{array}
$$

where the local gradient $\delta_j$ is defined as follows: (Seung, 2002):

1.  If node $j$ is an output node, then $\delta_j$ is the product of $\varphi'(v_j)$ and the error signal $e_j$, where $\varphi(\_)$ is the logistic function and $v_j$ is the total input to node $j$ (i.e. $\Sigma_i w_{ji} y_i$), and $e_j$ is the error signal for node $j$ (i.e. the difference between the desired output and the actual output);

2.  If node $j$ is a hidden node, then $\delta_j$ is the product of $\varphi'(v_j)$ and the weighted sum of the $\delta$'s computed for the nodes in the next hidden or output layer that are connected to node $j$.
    [The actual formula is $\delta_j = \varphi'(v_j)$ &Sigma$_k$ $\delta_k w_{kj}$ where $k$ ranges over those nodes for which $w_{kj}$ is non-zero (i.e. nodes $k$ that actually have connections from node $j$. The $\delta_k$ values have already been computed as they are in the output layer (or a layer closer to the output layer than node $j$).]

### 3.2 Analytic programming

Basic principles of the analytic programming (AP) were developed in 2001 (Zelinka, 2002). Until that time only genetic programming (GP) and grammatical evolution (GE) had existed. GP uses genetic algorithms while AP can be used with any evolutionary algorithm, independently on individual representation. To avoid any confusion, based on use of names according to the used algorithm, the name - Analytic Programming was chosen, since AP represents synthesis of analytical solution by means of evolutionary algorithms.

The core of AP is based on a special set of mathematical objects and operations. The set of mathematical objects is set of functions, operators and so-called terminals (as well as in GP), which are usually constants or independent variables. This set of variables is usually mixed together and consists of functions with different number of arguments. Because of a variability of the content of this set, it is called here "general functional set" – GFS. The structure of GFS is created by subsets of functions according to the number of their arguments. For example GFSall is a set of all functions, operators and terminals, GFS3arg is a subset containing functions with only three arguments, GFS0arg represents only terminals, etc. The subset structure presence in GFS is vitally important for AP. It is used to avoid synthesis of pathological programs, i.e. programs containing functions without arguments, etc. The content of GFS is dependent only on the user. Various functions and terminals can be mixed together (Zelinka, 2002; Oplatková, 2009).

The second part of the AP core is a sequence of mathematical operations, which are used for the program synthesis. These operations are used to transform an individual of a population into a suitable program. Mathematically stated, it is a mapping from an individual domain into a program domain. This mapping consists of two main parts. The first part is called discrete set handling (DSH), see Fig. 6 (Zelinka, 2002) and the second one stands for security procedures which do not allow synthesizing pathological programs. The method of DSH, when used, allows handling arbitrary objects including nonnumeric objects like linguistic terms {hot, cold, dark...}, logic terms (True, False) or other user defined functions. In the AP DSH is used to map an individual into GFS and together with security procedures creates the above mentioned mapping which transforms arbitrary individual into a program.

AP needs some evolutionary algorithm (Zelinka, 2004) that consists of population of individuals for its run. Individuals in the population consist of integer parameters, i.e. an individual is an integer index pointing into GFS. The creation of the program can be schematically observed in Fig. 7. The individual contains numbers which are indices into GFS. The detailed description is represented in (Zelinka, 2002; Oplatková, 2009).

AP exists in 3 versions – basic without constant estimation, APnf – estimation by means of nonlinear fitting package in *Mathematica* environment and APmeta – constant estimation by means of another evolutionary algorithms; meta means metaevolution.

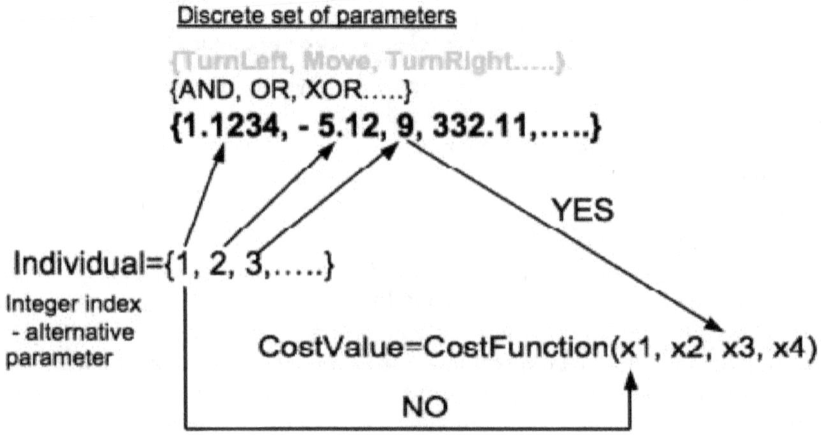

Discrete set of parameters

{TurnLeft, Move, TurnRight.....}
{AND, OR, XOR.....}
**{1.1234, - 5.12, 9, 332.11,.....}**

**YES**

Individual={1, 2, 3,.....}

Integer index
- alternative
parameter

CostValue=CostFunction(x1, x2, x3, x4)

**NO**

Fig. 6. Discrete set handling

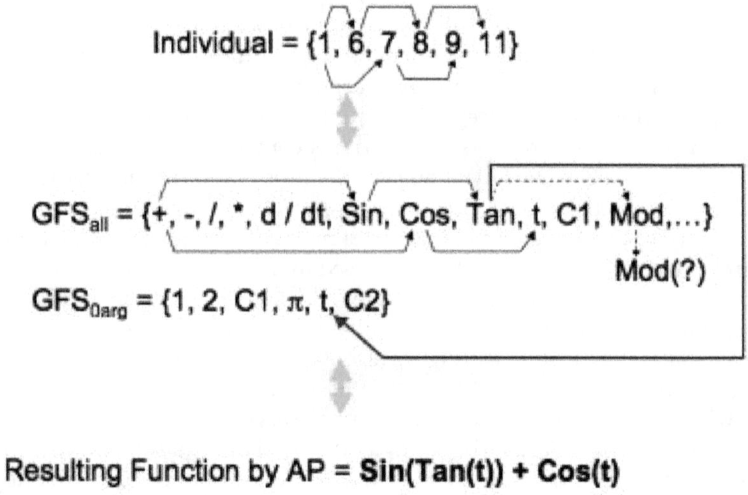

Individual = {1, 6, 7, 8, 9, 11}

GFS$_{all}$ = {+, -, /, *, d / dt, Sin, Cos, Tan, t, C1, Mod,...}

Mod(?)

GFS$_{0arg}$ = {1, 2, C1, π, t, C2}

Resulting Function by AP = **Sin(Tan(t)) + Cos(t)**

Fig. 7. Main principles of AP

## 4. Experimental results

### 4.1 Used datasets

This approach allows a search of structural shapes (patterns) inside time-series. Patterns are composed of simpler sub-patterns. The most elementary ones are known as primitives. Feature extraction is carried out by dividing the initial waveform into segments, which are encoded. Search for patterns is accomplished process, which is performed manually by the user. In order to test the efficiency of pattern recognition, we applied a database downloaded from (Google finance, 2010). We used time series, which shows development of the market value of U.S. company Google and represents the minute time series from 29 October 2010, see Fig. 8.

Used algorithms need for their adaptation training sets. In all experimental works, the training set consists of 100 samples (e.g. training pars of input and corresponding output vectors) and it is made from the time series and contains three peaks, which are indicated by vertical lines and they are shown in Figure 8. Samples obtained in this way are always adjusted for the needs of the specific algorithm. Data, which were tested in our experimental works, contains only one peak, which is indicated by vertical lines and it is shown in Fig. 9.

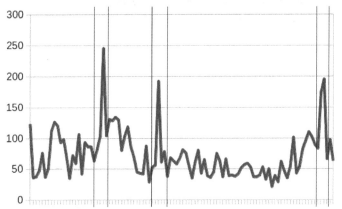

Fig. 8. The training set with three marked peaks

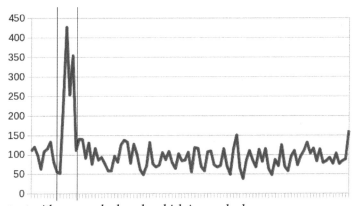

Fig. 9. The test set with one marked peak, which is searched

## 4.2 Pattern recognition via artificial neural networks

The aim of this experiment was to adapt neural network so that it could find one kind of pattern (peak) in the test data. We have used two sets of values, which are graphically depicted in Figure 10 (training patterns) and Figure 11 (test patterns) in our experiments. Training set always contained all define peaks, which were completed by four randomly

selected parts out of peaks. These randomly selected parts were used to network can learn
to recognize what is or what is not a search pattern (peak). All patterns were normalized to
the square of a bitmap of the edge of size $a = 10$. The effort is always to choose the size of
training set as small as possible, because especially backpropagation networks increases
their computational complexity with the size of a training set.

Fig. 10. Graphic representation of learning patterns (**S** vectors) that have been made by
selection from training data set. The first three patterns represent peaks. Next four patterns
are representatives of non-peak "not-interested" segments of values

| No. | S | T |
|-----|---|---|
| 0. | ----------+- \| ---------++- \| --------+++ \| -------++++ \| -------++++ \| <br> -----+++++ \| -----+++++ \| --+++++++++ \| +++++++++++ \| +++++++++++ | -+ |
| 1. | ---------- \| ---------- \| --------+- \| -------++- \| -------+++ \| <br> ------++++ \| ------++++ \| -----+++++ \| -----+++++ \| --++++++++ | -+ |
| 2. | ---------- \| ---------- \| --------++- \| -----++++- \| ----++++++ \| <br> ----++++++ \| ---+++++++ \| +++++++++++ \| +++++++++++ \| +++++++++++ | -+ |
| 3. | ---------- \| ---------- \| ---------- \| ---------- \| ---------- \| <br> ---------- \| ---------- \| ---------- \| -------+++ \| +++++++++++ | +- |
| 4. | ---------- \| ---------- \| ---------- \| ---------- \| ---------- \| <br> ---------- \| ---------- \| ---------- \| ---------- \| ++++++++++- | +- |
| 5. | ---------- \| ---------- \| ---------- \| ---------- \| ---------- \| <br> ---------- \| --------++ \| -------+++ \| ++-----++++ \| +++++++++++ | +- |
| 6. | ---------- \| ---------- \| ---------- \| ---------- \| ---------- \| <br> ---------- \| --------+- \| ------++++ \| --+++++++++ \| +++++++++++ | +- |

Table 1. Vectors **T** and **S** from the learning pattern set. Values of '-1' are written using the
character '-' and values of '+1' are written using the character '+' because of better clarity

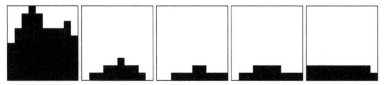

Fig. 11. Graphic representation of test patterns (S vectors) that have been made by selection from the test data set. The first pattern represents the peak. Next four patterns are representatives of non-peak "not-interested" segments of values

| No. | S | T |
|-----|---|---|
| 0. | `---+------\|--+++-----\|--+++---+-\|-++++++++-\|-+++++++++\|`<br>`++++++++++\|++++++++++\|++++++++++\|++++++++++\|++++++++++` | -+ |
| 1. | `----------\|----------\|----------\|----------\|----------\|`<br>`----------\|----------\|-----+----\|---+++++--\|-++++++++-` | +- |
| 2. | `----------\|----------\|----------\|----------\|----------\|`<br>`----------\|----------\|----------\|-----++---\|--+++++++++` | +- |
| 3. | `----------\|----------\|----------\|----------\|----------\|`<br>`----------\|----------\|----------\|---++++---\|-+++++++++` | +- |
| 4. | `----------\|----------\|----------\|----------\|----------\|`<br>`----------\|----------\|----------\|++++++++++-\|++++++++++` | +- |

Table 2. Vectors T and S from the test pattern set. Values of '-1' are written using the character '-' and values of '+1' are written using the character '+' because of better clarity

Two types of classifiers: Backpropagation and classifier based on Hebb learning were used in our experimental part. Both used networks classified input patterns into two classes. Backpropagation network was adapted according the training set (Fig.10, Tab. 1) in 7 cycles. After its adaptation, the network was able to also correctly classify all five patterns from the test set (Fig. 11, Tab. 2), e.g. the network was able to correctly identify the peak and "uninteresting" data segments too. Other experiments gave similar results too.

**Backpropagation network configuration:**

| | |
|---|---|
| Number of input neurons: | 100 |
| Number of output neurons: | 2 |
| Number of hidden layers: | 1 |

| Number of hidden neurons: | 3 |
| --- | --- |
| α - learning parameter: | 0.4 |
| Weight initialization algorithm: | Nguyen-Widrow |
| Weight initialization range: | (-0.5; +0.5) |
| Type of I/O values: | bipolar |

Hebb network in its basic configuration was not able to adapt given training set (Fig.10, Tab. 1), therefore we used modified version of the network removing useless components from input vectors (Kocian & Volná & Janošek & Kotyrba, 2011). Then, the modified Hebb network was able to adapt all training patters (Fig. 12) and in addition to that the network correctly classified all the patterns from the test set (Fig. 11, Tab. 2), e.g. the network was able to correctly identify the peak and "uninteresting" data segments too. Other experiments gave similar results too.

**Hebbian-learning-based-classifier configuration:**

| Number of input neurons: | 100 |
| --- | --- |
| Number of output neurons: | 2 |
| Type of I/O values: | bipolar |

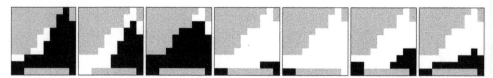

Fig. 12. Learning patterns from Fig. 10 with uncovered redundant components (gray colour). The redundant components prevented the Hebbian-learning-based-classifier in its default variant to learn patterns properly. So the modified variant had to be used

### 4.2 Pattern recognition via analytic programming

As an evolutionary algorithm used in our experimental work was differential evolution (DE). DE is a population-based optimization method that works on real-number-coded individuals (Price, 1999). For each individual $\vec{x}_{i,G}$ in the current generation (G), DE generates a new trial individual $\vec{x}'_{i,G}$ by adding the weighted difference between two randomly selected individuals $\vec{x}_{r1,G}$ and $\vec{x}_{r2,G}$ to a randomly selected third individual $\vec{x}_{r3,G}$. The resulting individual $\vec{x}'_{i,G}$ is crossed-over with the original individual $\vec{x}_{i,G}$. The fitness of the resulting individual, referred to as a perturbed vector $\vec{u}_{i,G+1}$, is then compared with the fitness of $\vec{x}_{i,G}$. If the fitness of $\vec{u}_{i,G+1}$ is greater than the fitness of $\vec{x}_{i,G}$, then $\vec{x}_{i,G}$ is replaced with $\vec{u}_{i,G+1}$; otherwise, $\vec{x}_{i,G}$ remains in the population as $\vec{x}_{i,G+1}$. DE is quite robust, fast, and effective, with global optimization ability. It does not require the objective function to be differentiable, and it works well even with noisy and time-dependent objective functions. The technique for the solving of this problem by means of analytic programming was inspired in neural networks. The method in this case study used input values and future output values – similarly as training set for the neural network and the whole structure which transfer input to output was synthesized by analytic programming. The final solution of the analytic programming is based on evolutionary process which selects only the required components from the basic sets of operators (Fig. 6 and Fig 7). Fig. 13 shows analytic programming experimental result for exact modelling during training phase.

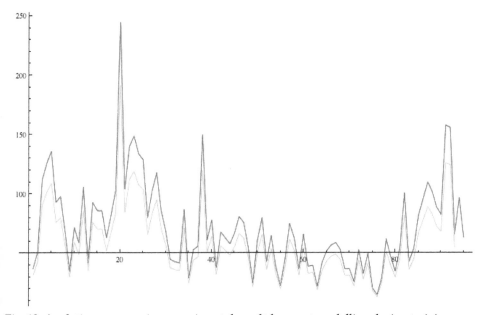

Fig. 13. Analytic programming experimental result for exact modelling during training phase. Red colour represents original data from training set (Fig. 8), while green colour represents modelling data using formula (6)

The resulting formula, which calculates the output value $x_n$ was developed using AP (6):

$$x_n = 85.999 \cdot e^{\left(17.1502 - x_{n-3}\right)^{0.010009 \cdot x_{n-2}}} \tag{6}$$

Analytic programming experimental results are shown in Fig. 14. Equation (6) also represents the behaviour of training set so that the given pattern was also successfully identified in the test set (Fig. 9). Other experiments gave similar results too. The operators used in GFS were (see Fig. 7): +, -, /, *, Sin, Cos, K, $x_{n-1}$ to $x_{n-4}$, exp, power. As the main algorithm for AP and also for constants estimation in meta-evolutionary process differential evolution was used. The final solution of the analytic programming is based on evolutionary process which selects only the required components from the basic sets of operators. In this case, not all components have to be selected as can be seen in one of solutions presented in (6).

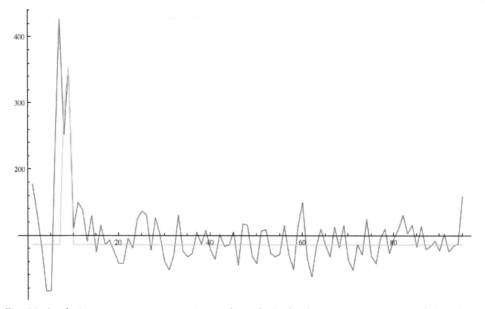

Fig. 14. Analytic programming experimental result. Red colour represents original data from test set (Fig. 9), while green colour represents modelling data using formula (6)

## 5. Conclusion

In this chapter, a short introduction into the field of pattern recognition using system adaptation, which is represented via time series, has been given. Two possible approaches were used from the framework of softcomputing methods. The first approach was based on analytic programming and the second one was based on artificial neural networks. Both types of used neural networks (e.g. Hebb and backpropagation networks) as well as analytic programming demonstrated ability to manage to learn and recognize given patterns in time series, which represents our system behaviour. Our experimental results suggest that for the given class of tasks can be acceptable simple classifiers (we tested the simplest type of Hebb learning). The advantage of simple neural networks is very easy implementation and quick adaptation. Easy implementation allows to realize them at low-performance computers (PLC) and their fast adaptation facilitates the process of testing and finding the appropriate type of network for the given application.

The method of analytic programming described here is universal (from point of view of used evolutionary algorithm), relatively simple, easy to implement and easy to use. Analytic programming can be regarded as an equivalent of genetic programming in program synthesis and new universal method, which can be used by arbitrary evolutionary algorithm. AP is also independent of computer platform (PC, Apple, ...) and operation system (Windows, Linux, Mac OS,...) because analytic programming can be realized for example in the Mathematica® environment or in other computer languages. It allows manipulation with symbolic terms and final programs are synthesised by AP of mapping, therefore main benefit of analytic programming is the fact that symbolic regression can be done by arbitrary evolutionary algorithm, as was proofed by comparative study.

According to the results of experimental studies, it can be stated that pattern recognition in our system behaviour using all presented methods was successful. It is not possible to say with certainty, which of them reaches the better results, whether neural networks or analytic programming. Both approaches have an important role in the tasks of pattern recognition.

In the future, we would like to apply pattern recognition tasks with the followed system adaptation methods in SIMATIC environment. SIMATIC (SIMATIC, 2010) is an appropriate application environment for industrial control and automation. SIMATIC platform can be applied at the operational, management and the lowest, physical level. At an operational level, it particularly works as a control of the running processes and monitoring of the production. On the management and physical level it can be used to receive any production instructions from the MES system (Manufacturing Execution System - the corporate ERP system set between customers' orders and manufacturing systems, lines and robots). At the physical level it is mainly used as links among various sensors and actuators, which are physically involved in the production process (Janošek, 2010). The core consists of the SIMATIC programmable logic computers with sensors and actuators. This system collects information about its surroundings through sensors. Data from the sensors can be provided (e.g. via Ethernet) to proposed and created software tools for pattern recognition in real time, which runs on a powerful computer.

## 6. Acknowledgment

The research described here has been financially supported by University of Ostrava grant SGS23/PRF/2011. It was also supported by the grant NO. MSM 7088352101 of the Ministry of Education of the Czech Republic, by grant of Grant Agency of Czech Republic GACR 102/09/1680 and by the European Regional Development Fund under the Project CEBIA-Tech No. CZ.1.05/2.1.00/03.0089. Any opinions, findings and conclusions or recommendations expressed in this material are those of the authors and do not necessarily reflect the views of the sponsors.

## 7. References

Armstrong, M. and Porter, R. ed. (2006): *Handbook of Industrial Organization, vol. III.* New York and Amsterdam: North-Holland.

Bishop, C. (2006) *Pattern Recognition and Machine Learning.* Springer, 2006.

Buhr, R.J.A. and Hubbard, A. (1997) Use Case Maps for Engineering Real Time and Distributed Computer Systems: A Case Study of an ACE-Framework Application. In *Hawaii International Conference on System Sciences*, Jan 7-10, 1997, Wailea, Hawaii, Available from http://www.sce.carletonca/ftp/pub/UseCaseMaps/hicss-final-public.ps

Ciskowski, P. and Zaton, M. (2010) Neural Pattern Recognition with Self-organizing Maps for Efficient Processing of Forex Market Data Streams. In *Artificial Intelligence and Soft Computing*, Volume 6113/2010, pp. 307-314, DOI: 10.1007/978-3-642-13208-7_39

Dormido-Canto, S., Farias, G., Vega, J., Dormido, R., Sánchez, J. and N. Duro et al. (2006) *Rev. Sci. Instrum.* 77 (10), p. F514.

Fausett, L.V. (1994) *Fundamentals of neural networks: architectures, algorithms and applications, first edition.* Prentice Hall. ISBN 978-953-7619-24-4

Gershenson, C. ( 2007): *Design and Control of Self-organizing Systems*. Mexico: CopIt ArXives, ISBN: 978-0-9831172-3-0.

Gershenson, C. (2002) Complex philosophy. In: *Proceedings of the 1st Biennial Seminar on Philosophical, Methodological & Epistemological Implications of Complexity Theory*. La Habana, Cuba. 14.02.2011, Available from http://uk.arXiv.org/abs/nlin.AO/0108001

Gogle finance [online], http://www.google.com/finance?q=NASDAQ:GOOG, 10.8. 2010

Heylighen, F. (1994) Fitness as default: the evolutionary basis for cognitive complexity reduction. In Trappl (Ed.) *Proceedings of Cybernetics and Systems '94*, R. Singapore: World Science, pp. 1595–1602, 1994.

Janošek, M. (2010) Systémy Simatic a jejich využití ve výzkumu. In: *Studentská vědecká konference 2010*. Ostrava: Ostravská univerzita, pp. 177-180. ISBN 978-80-7368-719-9

Kocian, V., Volná, E., Janošek, M. and Kotyrba, M. (2011) Optimizatinon of training sets for Hebbian-learningbased classifiers. In R. Matoušek (ed.): *Proceedings of the 17th International Conference on Soft Computing, Mendel 2011*, Brno, Czech Republic, pp. 185-190. ISBN 978-80-214-4302-0, ISSN 1803-3814.

Lai, K.K., Yu, L. and Wang, S: A (2005) Neural Network and Web-Based Decision Support System for Forex Forecasting and Trading. In *Data Mining and Knowledge Management*, Volume 3327/2005, pp. 243-253, DOI: 10.1007/978-3-540-30537-8_27.

Oplatkova, Z. (2009) Metaevolution - Synthesis of Optimization Algorithms by means of Symbolic. In *Regression and Evolutionary Algorithms, Lambert-Publishing*, ISBN 978-8383-1808-0.

Price, K. (1999) An Introduction to Differential Evolution, In: (D. Corne, M. Dorigo and F. Glover, eds.) *New Ideas in Optimization*, pp. 79–108, London: McGraw-Hill.

Seung, S. (2002). Multilayer perceptrons and backpropagation learning. 9.641 Lecture4. 1-6. Available from:
http://hebb.mit.edu/courses/9.641/2002/lectures/lecture04.pdf

SIMATIC (2010) [online]. SIMATIC Controller, Available from
http://www.automation.siemens.com/salesmaterial-as/brochure/en/brochure_simatic-controller_en.pdf

Zelinka, I. (2002) Analytic programming by Means of Soma Algorithm. Mendel '02, In: *Proc. Mendel'02*, Brno, Czech Republic, 2002, 93-101., ISBN 80-214-2135-5

Zelinka, I. (2004) SOMA – Self Organizing Migrating Algorithm", In: B.V. Babu, G. Onwubolu (eds), *New Optimization Techniques in Engineering Springer-Verlag*, 2004, ISBN 3-540-20167X

# 10

# Sequential and Simultaneous Algorithms to Solve the Collision-Free Trajectory Planning Problem for Industrial Robots – Impact of Interpolation Functions and the Characteristics of the Actuators on Robot Performance

Francisco J. Rubio, Francisco J. Valero, Antonio J. Besa and Ana M. Pedrosa
*Centro de Investigación de Tecnología de Vehículos, Universitat Politècnica de València*
*Spain*

## 1. Introduction

Trajectory planning for robots is a very important issue in those industrial activities which have been automated. The introduction of robots into industry seeks to upgrade not only the standards of quality but also productivity as the working time is increased and the useless or wasted time is reduced. Therefore, trajectory planning has an important role to play in achieving these objectives (the motion of robot arms will have an influence on the work done).

Formally, the trajectory planning problem aims to find the force inputs (control $u(t)$) to move the actuators so that the robot follows a trajectory $q(t)$ that enables it to go from the initial configuration to the final one while avoiding obstacles. This is also known as the complete motion planning problem compared with the path planning problem in which the temporal evolution of motion is neglected.

An important part of obtaining an efficient trajectory plan lies with both the interpolation function used to help obtain the trajectory and the robot actuators. Ultimately actuators will generate the robot motion, and it is very important for robot behavior to be smooth. Therefore, the trajectory planning algorithms should take into account the characteristics of the actuators without forgetting the interpolation functions which also have an impact on the resulting motion. As well as smooth robot motion, it is also necessary to monitor some working parameters to verify the efficiency of the process, because most of the time the user seeks to optimize certain objective functions. Among the most important working parameters and variables are the time required to get the trajectory done, the input torques, the energy consumed and the power transmitted. The kinematic properties of the robot´s links, such as the velocities, accelerations and jerks are also important.

The trajectory algorithm should also not overlook the presence of possible obstacles in the workspace. Therefore it is very important to model both the workspace and the obstacles efficiently. The quality of the collision avoidance procedure will depend on this modelization.

## 2. A brief look at previous work

Trajectory planning for industrial robots is a very important topic in the field of robotics and has attracted a great number of researchers so that there are at the moment a variety of methodologies for its resolution.

By studying the work done by other researchers on this topic it is easy to deduce that the problem has mainly been tackled with two different approaches: direct and indirect methods. Some authors who have analyzed this topic using indirect methods are Saramago, 2001; Valero et al., 2006; Gasparetto and Zanotto, 2007 ; du Plessis et al., 2003.

Other authors, on the other hand, have implemented the direct method such as Chettibi et al, 2002; Macfarlane, 2003; Abdel-Malek et al. 2006. However, in these examples the obstacles have been neglected which is a drawback.

Over the years, the algorithms have been improved and the study of the robotic system has become more and more realistic. One way of achieving that is to analyze the complete behavior of the robotic system, which in turn leads us to optimize some of the working parameters mentioned earlier by means of the appropriate objective functions. The most widely used optimization criteria can be classified as follows:

1.    Minimum time required, which is bounded to productivity.
2.    Minimum jerk, which is bounded to the quality of work, accuracy and equipment maintenance.
3.    Minimum energy consumed or minimum actuator effort, both linked to savings.
4.    Hybrid criteria, e.g. minimum time and energy.

The early algorithms that solved the trajectory planning problem tried to minimize the time needed for performing the task (see Bobrow et al., 1985; Shin et al., 1985; Chen et al., 1989). In those studies, the authors impose smooth trajectories to be followed, such as spline functions.

Another way of tackling the trajectory planning problem was based on searching for jerk-optimal trajectories. Jerks are essential for working with precision and without vibrations. They also affects the control system and the wearing of joints and bars. Jerk constraints were introduced by Kyriakopoulos (see Kyriakopoulos et al.,1988). Later, Constantinescou introduces (Constantinescu et all, 2000) a method for determining smooth and time-optimal path-constrained trajectories for robotic manipulators imposing limits on the actuator jerks.

Another different approach to solving the trajectory planning problem is based on minimizing the torque and the energy consumed instead of the execution time or the jerk. An early example is seen in Garg et al., 1992. Similarly, Hirakawa and Kawamura searched for the minimum energy consumed (Hirakawa et al., 1996). In Field and Stepanenko, 1996, the authors plan minimum energy consumption trajectories for robotic manipulators. In Saramago and Steffen, 2000, the authors considered not only the minimum time but also the minimum mechanical energy of the actuators. They built a multi-objective function and the results obtained depended on the associated weighting factor. The subject of energy minimization continues to be of interest in the field of robotics and automated manufacturing processes ( Cho et al., 2006 ).

Later, new approaches appear for solving the trajectory planning problem. The idea of using a weighted objective function to optimize the operating parameters of the robot arises (Chettibi et al., 2004). Gasparetto and Zanotto also use a weighted objective function (see Gasparetto and Zanotto, 2010). In this chapter we will introduce an indirect method which has been called the "sequential algorithm".

In this chapter we will describe two algorithms for solving the collision-free trajectory planning for industrial robots that we have developed. We have called them "sequential" and "simultaneous" algorithms. The first is an indirect method while the second is a direct one. The "sequential" algorithm considers the main properties of the actuators (torque, power, jerk and energy consumed). The "simultaneous" algorithm analyzes what is the best interpolation function to be used to generate the trajectory considering a simple actuator (only the torque required). The chapter content is based on the previous work done by the authors (see Valero et al., 2006, and Rubio et al., 2007). Specifically, the two approaches to solving the trajectory planning problem are explained.

## 3. Robot modelling

The robot model used henceforth is the wire model corresponding to the PUMA 560 robot shown in Fig. 1. The robot involves rigid links that are joined by the corresponding kinematic joints (revolution).The robot has $F$ degrees of freedom and each robot´s configuration $C^j$ can be unequivocally set using the Cartesian coordinates of N points, which are called significant points. These points, defined as $\alpha^j_i$ ($x^j_{3(i-1)+1}$ ,$x^j_{3(i-1)+2}$ , $x^j_{3(i-1)+3}$, $i=1..F$, $j$=number of configuration) are chosen systematically. Therefore, ultimately, every configuration will be expressed in Cartesian coordinates by means of the significant points, i.e. $C^j = C^j (\alpha^j_i)$, which represent the specifics of the robot under study. It is important to point out that they do not constitute an independent set of coordinates. Besides the significant points, some other points $p^j_k$, called interesting points, will be used to improve the efficiency of the algorithms, the coordinates of which are obtained from the significant points and the geometric characteristics of the robot.

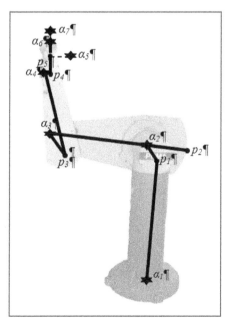

Fig. 1. Model of robot PUMA 560. Significant and Interesting Points (mobile base)

The PUMA 560 robot can be modelled with a movable base or a fixed base. The mobile-based robot is shown in Fig. 1 with the seven significant points used ($\alpha_1$, $\alpha_2$, $\alpha_3$, $\alpha_4$, $\alpha_5$, $\alpha_6$ and $\alpha_7$ ), together with another five interesting points $p$ $j_k$. As a result of this, the configuration $C^j$ is determined by twenty-one variables corresponding to the coordinates of the significant points. These variables are connected through fourteen constraint equations relative to the geometric characteristics of the robot (length of links, geometric constraints and range of motion). See Rubio et al, 2009 for more details. It must be noted that any other industrial robot can be modelled in this way by just selecting and choosing appropriately those significant points that best describe it.

This property is very important as far as the effectiveness of the algorithm is concerned.

## 4. Workspace modelling

The workspace is modelled as a rectangular prism with its edges parallel to the axes of the Cartesian reference system. The work environment is defined by the obstacles bound to produce collisions when the robot moves within the workspace. The obstacles are considered static, i.e. their positions do not vary over time and they are represented by means of unions of patterned obstacles.

The fact of working with patterned obstacles introduces two fundamental advantages:
1. It allows the modelling of any generic obstacle so that collisions with the robot´s links can be avoided.
2. It permits working with a reduced number of patterned obstacles in order to model a complex geometric environment so that its use is efficient. It means that a small number of constraints are introduced into the optimization problem when obtaining collision-free adjacent configurations.

The patterned obstacles have a geometry based on simple three-dimensional figures, particularly spheres, rectangular prisms and cylinders. Any obstacle present in the workspace could be represented as a combination of these geometric figures.

The definition of a patterned obstacle is made in the following way:
- Spherical obstacle $SO_i$, is defined when the position of its centre and its radius are known. It is characterized by means of

- Centre of Sphere $i$: $c_i^{SO} = \left( c_{xi}^{SO}, c_{yi}^{SO}, c_{zi}^{SO} \right)$

- Radius of Sphere $i$: $r_i^{SO}$

Therefore $SO_i = \left( c_i^{SO}, r_i^{SO} \right)$. See Fig. 2

Fig. 2. Generic Spherical obstacle $SO_i$

- Cylindrical obstacle $CO_k$, is defined when the coordinates of the centres of its bases and its radius are known. It is characterized by means of
- Centre of base 1 for cylinder $k$: $c_{1k}^{CO} = \left( c_{1xk}^{CO}, c_{1yk}^{CO}, c_{1zk}^{CO} \right)$
- Centre of base 2 for cylinder $k$: $c_{2k}^{CO} = \left( c_{2xk}^{CO}, c_{2yk}^{CO}, c_{2zk}^{CO} \right)$
- Radius of cylinder $k$: $r_k^{CO}$

Therefore $CO_k = \left( c_{1k}^{CO}, c_{2k}^{CO}, r_k^{CO} \right)$. See Fig. 3

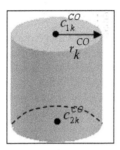

Fig. 3. Generic Cylindrical obstacle $CO_k$

- Prismatic obstacle $PO_l$, is defined when four points located in the vertices of the rectangular prism are known so that vectors that are perpendicular to each other can be drawn up. It is characterized by means of
- Point $a$ of prism l: $a_l^{PO} = \left( a_{xl}^{PO}, a_{yl}^{PO}, a_{zl}^{PO} \right)$
- Point $q_1$ of prism l: $q_{1l}^{PO} = \left( q_{1xl}^{PO}, q_{1yl}^{PO}, q_{1zl}^{PO} \right)$
- Point $q_2$ of prism l: $q_{2l}^{PO} = \left( q_{2xl}^{PO}, q_{2yl}^{PO}, q_{2zl}^{PO} \right)$
- Point $q_3$ of prism l: $q_{3l}^{PO} = \left( q_{3xl}^{PO}, q_{3yl}^{PO}, q_{3zl}^{PO} \right)$

Therefore $PO_l = \left( a_l^{PO}, q_{1l}^{PO}, q_{2l}^{PO}, q_{3l}^{PO} \right)$. See Fig. 4

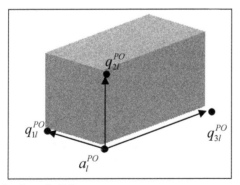

Fig. 4. Generic Prismatic obstacle $PO_l$

## 5. Discretizing the workspace

With the purpose of working with a limited number of configurations, the generation of a discrete workspace that represents the possible positions of the end-effector of the robot is considered. To do this, a rectangular prism with its edges parallel to the axes of the

Cartesian reference system is created and whose opposite vertices correspond to the positions of the end-effector of the robot in the initial and final configurations from which the connecting path is calculated. The set of positions that the end-effector of the robot can adopt within the prism is restricted to a finite number of points resulting from the discretization of the prism according to the following increases:

$$\Delta x = \frac{\left|\alpha_{nx}^{f} - \alpha_{nx}^{i}\right|}{N_x} \quad \Delta y = \frac{\left|\alpha_{ny}^{f} - \alpha_{ny}^{i}\right|}{N_y} \quad \Delta z = \frac{\left|\alpha_{nz}^{f} - \alpha_{nz}^{i}\right|}{N_z} \tag{1}$$

Where the values of $\Delta x$, $\Delta y$ and $\Delta z$ are calculated from the values of the number of intervals $N_x$, $N_y$ and $N_z$ in which the prism is discretized, and those increments should be smaller than the smallest dimension of the obstacle modelled in the workspace. Points ($\alpha_{nx}^{f}$, $\alpha_{ny}^{f}$, $\alpha_{nx}^{f}$) and ($\alpha_{nx}^{i}$, $\alpha_{ny}^{i}$, $\alpha_{nz}^{i}$) correspond to the coordinates of the end-effector of the robot for the initial and final configurations. Fig. 6 demonstrates the way in which the prism that gives rise to the set of nodes that the end-effector of the PUMA 560 robot with a mobile base can adopt is discretized.

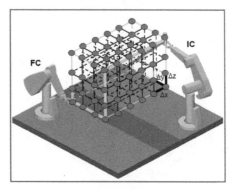

Fig. 5. Rectangular prism with edges parallel to the axes of the Cartesian reference system

## 6. Obstacle avoidance
By controlling the distance from the different patterned obstacles to the cylinders that cover the robot links, collision avoidance between the robot and the obstacles is possible. Distances are constraints in the optimization problem. They serve to calculate collision-free adjacent configurations (for adjacent configuration see Section 7).

### 6.1 Calculation of distances
Each robot's link is modelled as a cylinder and it is characterized as $RC_i = \left(c_{1i}^{RC}, c_{2i}^{RC}, r_i^{RC}\right)$ (see section 4). The application of the procedure to calculate distances between link $i$ of the robot and the patterned obstacle $j$ (which may be a cylinder, sphere or a prism), can give rise to three different cases to prevent collisions:

### A) Cylinder-Sphere

See Fig. 6. Here we compute the distance between a line segment (cylinder) to a point (centre of the sphere). Let $AB$ be a line segment specified by the endpoints $A$ and $B$. Given an arbitrary point $C$, the problem is to determine the point $P$ on $AB$ closest to $C$. Then we calculate the distance between these two points.

Projecting $C$ onto the extended line through $AB$ provides the solution. If the projection point $P$ lies within the segment, then $P$ itself is the correct answer. If $P$ lies outside the segment, then the segment endpoint closest to $C$ is instead the closest point (A or B).

## B) Cylinder-Cylinder

See Fig. 6. Here we compute the distance between two line segments. The problem of determining the closest points between two line segments $S_1$ ($P_1Q_1$) and $S_2$ ($P_2Q_2$) (and therefore the distance) is more complicated than computing the closest points of the lines $L_1$ and $L_2$ of which the segments are a part. Only when the closest points of $L_1$ and $L_2$ happen to lie on the segments does the method for closest points between lines apply. For the case in which the closest points between $L_1$ and $L_2$ lie outside one or both segments, a common misconception is that it is sufficient to clamp the outside points to the nearest segment endpoint. It can be shown that if just one of the closest points between the lines is outside its corresponding segment, that point can be clamped to the appropriate endpoint of the segment and the point on the other segment closest to the endpoint is computed. If both points are outside their respective segments, the same clamping procedure must be repeated twice.

## C) Cylinder-Prism

See Fig. 6. The prismatic surfaces are divided into triangles. In this case we compute the distance between a line segment and a triangle. The closest pair of points between a line segment $PQ$ and a triangle is not necessarily unique. When the line segment is parallel to the plane of the triangle, there may be an infinite number of equally close pairs. However, regardless of whether the segment is parallel to the plane or not, it is always possible to locate a point such that the minimum distance falls either between the end point of the segment and the interior of the triangle or between the segment and an edge of the triangle. Thus, the closest pair of points (and therefore the distance) can be found by computing the closest pairs of points between the following entities:

- Segment $PQ$ and triangle edge $AB$.
- Segment $PQ$ and triangle edge $BC$.
- Segment $PQ$ and triangle edge $CA$.
- Segment endpoint $P$ and plane of triangle (when $P$ projects inside $ABC$)
- Segment endpoint $Q$ and plane of triangle (when $Q$ projects inside $ABC$).

The number of tests required to calculate the distance can be reduced in some cases.

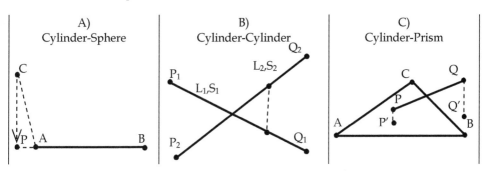

Fig. 6. Three different cases to calculate distances (and prevent collisions)

## 7. Obtaining adjacent configurations

The discrete configuration space is obtained by means of generating adjacent configurations. Given a feasible configuration $C^k$, it is said that a new configuration $C^p$ is adjacent to the first if it is also feasible (i.e. it fulfils the characteristics associated to the robot modelling and avoids collisions with the obstacles), and in addition the following three properties are fulfilled:

1. The position of the end-effector that corresponds to a node of the discrete workspace is at a distance of one unit with respect to the position of the end-effector of configuration $C^k$. That means that at least one of the following conditions has to be fulfilled:

$$\left|\alpha_{nx}^k - \alpha_{nx}^p\right| = \Delta x \qquad \left|\alpha_{ny}^k - \alpha_{ny}^p\right| = \Delta y \qquad \left|\alpha_{nz}^k - \alpha_{nz}^p\right| = \Delta z \tag{2}$$

$n$ being the subscript corresponding to the significant point associated to the end-effector of the robot. For PUMA 560 with mobility in the base, $n=7$, as can be seen in Fig. 1.
What we obtain is a sequence of configurations that is contained in the path, so that by using interpolation we can obtain a collision-free and continuous path.

2. Verification of the absence of obstacles between adjacent configurations $C^k$ and $C^p$. Since the algorithm works in a discrete space it is necessary to verify that there are no obstacles between adjacent configurations, for which the following condition is set out:

$$\left|\overrightarrow{\alpha_i^k \alpha_i^p}\right| \le 2 \cdot min\left(r_j\right) \tag{3}$$

where $r_j$ is the characteristic dimension of the smallest patterned obstacle. This condition is necessary to guarantee that the distance for each link between two adjacent configurations is less than the characteristic dimension of the smallest patterned obstacle.

3. Configuration $C^p$ must be such that it minimizes the following objective function:

$$\left\|C^p - C^f\right\| = \sum_{i=1}^{n}\left(\left(\alpha_{xi}^p - \alpha_{xi}^f\right)^2 + \left(\alpha_{yi}^p - \alpha_{yi}^f\right)^2 + \left(\alpha_{zi}^p - \alpha_{zi}^f\right)^2\right) \tag{4}$$

$n$ being the number of significant points of the robot. This third property facilitates the final configuration to be reached even for redundant robots, i.e. the robot's end-effector should not be part, at the final node, of a configuration different from the desired one. On the other hand, this property has an influence on the configurations generated, facilitating the configurations in the neighbourhoods at the end so that they are compatible with the end.

An optimization procedure is set by using a sequential quadratic programming method (SQP). This method serves to minimize a quadratic objective function subject to a set of constraints which might include those from a simple limit to the values of the variables, linear restrictions and nonlinear continuous constraints. This is an iterative method.

Applying this procedure to the path planning problem, the objective function used is given by Eq. (4). The constraints are associated to the geometry of the robot, the limits of the actuators and the avoidance of collision. And the configuration $C^k$ is used as an initial estimation for its resolution. The solution of this optimization problem gives the adjacent configuration $C^p$ looked for. By repeating the obtaining of adjacent configuration, the discrete configuration space of the robot is obtained. These configurations are recorded in a graph.

## 8. "Sequential" algorithm applied to solving the trajectory planning problem. Problem statement

### 8.1 Introduction

The "sequential" algorithm is based on an indirect approach to solving the trajectory planning problem. The algorithm takes into account the characteristics of the actuators (torque, power, jerk and consumed energy), the interpolation functions and the obstacles in the workspace. It generates the configuration space. Then, a graph is associated to the previously obtained configuration space, which allows a collision-free path to be obtained between the initial and final configurations. Once the path is available, the dynamic characteristics of the robot are included, setting an optimal trajectory planning problem which aims to obtain the minimum time trajectory that is compatible with the robot features and the actuator capabilities (torque, jerk and consumed energy constraints).

### 8.2 Obtaining a path

First, the algorithm solves the path planning problem, obtaining the discrete configuration space of the robot (the discrete configuration space is generated by means of adjacent configurations, see Section 7) and then the minimum distance path is calculated. This path (a sequence of $m$ configurations) is obtained by associating a weighted graph to the discrete configuration space and looking for the minimum weighted path. In the graph, the nodes correspond to the robot configurations and the arcs are related to joint displacements between adjacent configurations.

The weight corresponding to the arc that goes from node $k$ ( $C^k$ robot configuration) to node $p$ ( $C^p$ robot configuration), can be given as:

$$a(k,p) = \sum_{i=1}^{3(F-1)} \left( x_i^p - x_i^k \right)^2 \tag{5}$$

when that $C^k$ and $C^p$ are adjacent. In addition $C^k$ and $C^p$ must satisfy both type (3) constraints that avoid the obstacles between configurations and the angle increased from $C^k$ to $C^p$ must be smaller than the magnitude of the forbidden zone for that joint, so that large displacements are avoided for movement between adjacent configurations.

In case the points above mentioned are not satisfied, then we consider that $a(k,p) = \infty$. Finally, the searching is started in the weighed graph with the path that joins the node corresponding to the initial configuration to the node corresponding to the final configuration. Since the arcs satisfy that $a(k,p) \geq 0$, the Dijkstra's algorithm is used to obtain the path that minimises the distance between the initial and final configurations. If this path exists, it is easy to obtain a sequence of $m$ robot configurations $S$.

### 8.3 Interpolation function

Once the path has been obtained (at this point, the algorithm uses Cartesian coordinates), we have a sequence of $m$ robot configurations, $S = \{S_1(q_{i1}), S_2(q_{i2})... S_m(q_{im})\}$. These configurations are expressed now in joint coordinates. And the objective now is to look for a minimum time trajectory ($t_{min}$), that contains them. The path is decomposed into $m$-1 intervals, so the time needed to reach the $S_{j+1}$ configuration from the initial $S_1$ is $t_j$, and the time spent in the segment $j$ (between $S_j$ and $S_{j+1}$ configurations) will be $t_j$-$t_{j-1}$. Cubic interpolation functions have been used for joint trajectories. They are defined by means of joint variables between successive configurations, so that for the segment $j$ it is:

$\forall t \in \left[ t_{j-1}, t_j \right[ \Rightarrow q_{ij} = a_{ij} + b_{ij}t + c_{ij}t^2 + d_{ij}t^3$ for $i=1,\ldots,dof$ ($dof$ being the degrees of freedoom of the robot) and $j=1,\ldots,m$-1. ($m$ is the number of the robot configuration)

To ensure motion continuity between configurations, the following conditions associated to the given configurations are considered.

• Position: it gives a total of ($2dof\,(m$-$1)$) equations:

$$q_{ij}\left( t_{j-1} \right) = a_{ij} + b_{ij}t_{j-1} + c_{ij}t_{j-1}^2 + d_{ij}t_{j-1}^3 \tag{6}$$

$$q_{ij}\left( t_j \right) = a_{ij} + b_{ij}t_j + c_{ij}t_j^2 + d_{ij}t_j^3 \tag{7}$$

• Velocity: for each interval, the initial and final velocity is zero, the velocity condition gives place to ($2dof$) equations:

$$\dot{q}_{i1}\left( t_0 \right) = 0 \tag{8}$$

$$\dot{q}_{im}\left( t_m \right) = 0 \tag{9}$$

When passing through each configuration, the final velocity of the previous configuration should be equal to the initial velocity of the next configuration, leading to ($dof\,(m$-$2)$) equations

$$\dot{q}_{ij}\left( t_j \right) = \dot{q}_{ij+1}\left( t_j \right) \tag{10}$$

• Acceleration: For each intermediate configuration, the final acceleration of the previous configuration should be equal to the initial acceleration of the next configuration, giving rise to ($dof(m$-$2)$) equations:

$$\ddot{q}_{ij}\left( t_j \right) = \ddot{q}_{ij+1}\left( t_j \right) \tag{11}$$

In addition, the minimum time trajectory must meet the following constraints:

• Maximum torque on the actuators,

$$\tau_i^{\min} \le \tau_i(t) \le \tau_i^{\max} \quad \forall t \in \left[0, t_{\min}\right], i = 1\ldots dof \tag{12}$$

• Maximum power on the actuators,

$$P_i^{\min} \le P_i(t) \le P_i^{\max} \quad \forall t \in \left[0, t_{\min}\right], i = 1\ldots dof \tag{13}$$

• Maximum jerk on the actuators,

$$\dddot{q}_i^{\min} \le \dddot{q}_i(t) \le \dddot{q}_i^{\max} \quad \forall t \in \left[0, t_{\min}\right], \ i = 1\ldots dof \tag{14}$$

• Consumed Energy,

$$\sum_{j=1}^{m-1} \left( \sum_{i=1}^{dof} \varepsilon_{ij} \right) \le E \ , \tag{15}$$

$\varepsilon_{ij}$ being the energy consumed by the $i$ actuator between configurations $j$ and $j+1$

Given the large number of iterations required by the process, the technique used for obtaining the coefficients is crucial. The first task is to normalize the polynomials that define the stages (see Suñer et al., 2007).In short, the optimization problem is set by using incremental time variables in each interval, so that in the interval between $S_j$ and $S_{j+1}$, the time variable should be $\Delta t_j = t_j - t_{j-1}$, and the objective function,

$$\sum_{j=1}^{m-1} \Delta t_j = t_{min} \qquad (16)$$

The solution is obtained by means of SQP procedures, so that at each iterative step it is necessary to obtain the above mentioned polynomials coefficients from the estimation of the variables of the problem.

## 8.4 Obtaining a trajectory

The trajectory is obtained when the optimization problem posed has been solved. The solution (and therefore the trajectory) is achieved by solving an optimization problem whose objective function is the trajectory total time and the constraints are the maximum torques in the robot actuators, maximum power, maximum jerk and the consumed energy. The solution of the optimization problem is approached by means of a SQP algorithm of Fortran mathematical library NAG. In each iterative step is necessary  to obtain the coefficients of the previously mentioned polynomials from an estimation of the variables $(t_j)$. Notice that the previous conditions above mentioned   define a system of $(4N_{dof}(m-1))$ independent linear equations. Since the complete trajectory has  $(4N_{dof}(m-1))$ unknowns corresponding to the coefficients of the polynomials, the linear system can be solved, obtaining the complete trajectory. And this linear system is solved in each iteration whithin the optimization problem. These coefficients are necessary to calculate the maximum torque, power, jerk and consumed energy for each one of the actuators by means of solving the inverse dynamic problem in each interval.

Finally, when the optimization problem has been solved we obtain the minimum time trayectory (subject to the mentioned constraints) and also all the kinematic properties of the robotic system.

## 8.5 Impact of interpolation function

The impact of the interpolation function is very important from the point of view of the robot´s performance. Polynomial interpolation functions have been used in the "sequential" algorithm. It has been noticed during the resolution of the examples that they extract the maximum dynamic capabilities of the robot´s actuators, so that the robot moves faster than if any other interpolation function is used (harmonic functions, etc). Therefore when the polynomial interpolation functions are used the algorithm gives the best results from the point of view of the time requiered to do the tasks.

## 8.6 Application and examples solved

Different examples have been solved for a PUMA 560 robot. The examples have been solved with sequences of different initial and final configurations. The trajectories calculated meet constraints on torque, power, jerk and energy consumed and the goal is to analyze impact of these constraints on the generation of minimum time collision-free trajectories for industrial robots. The results obtained show that constraints on the energy consumed must enable the manipulator to exceed the requirements associated with potential energy, as the algorithm works on the assumption that the energy can be dissipated but not recovered. Also, an increase in the severity of energy constraints results in longer time trajectories with more soft power requirements. When constraints are not very severe, efficient trajectories can be obtained without high penalties on the working time cycle. An increase in the severity of the jerk constraints involves longer time trajectories with more soft power requirements and lower energy consumed. When constraints are very severe, times are also severely penalized even the jerk might appear. To obtain competitive results in the balance between time cycle and energy consumed, the actuators should work with the maximum admissible value of the jerk so that the robot can work with the desired accuracy.

## 9. "Simultaneous" algorithm applied to solving the trajectory planning problem. Problem statement

### 9.1 Introduction
The "simultaneous" algorithm is based on a direct approach to solving the trajectory planning problem in which the path planning problem and the problem of determining the time history of motion are treated as one instead of treating them separately as the indirect methods do. The algorithm is called "simultaneous" because of the simultaneous generation of discrete configuration space and the minimum distance path, making use of the information that the objective function is generating when new configurations are obtained. The algorithm works on a discretized configuration space which is generated gradually as the direct procedure solution evolves. It uses Cartesian coordinates (to specify the motion of the end-effector) and joint coordinates (to solve the inverse dynamic problem). An important role is played by the generation of adjacent configurations using techniques described by Valero et al. ,2006 . The resolution of the inverse dynamic problem has been done using Gibbs-Appel´s equations, as proposed by Sebastian Provenzano ( see Provenzano, 2001). Any obstacle can be modelled using simple obstacle patterns: sphere, cylinder and prism. This helps calculate distances and avoid collisions.

The algorithm takes into account the torque required by the actuators, analyses the best interpolation function and consider the obstacles in the workspace. To obtain a new adjacent configuration $C^k$ , a first optimization problem has to be solved which can be stated as follows:

$$\text{Find } C^k, \text{ minimizing } Min\left(\left\|C^k - C^f\right\|\right) = Min\left(\sqrt{\sum_{i=1}^{n}\left(x_i^k - x_i^f\right)}\right) \tag{17}$$

and subject to:
a.   Geometrical constraints of the robot structure;
b.   Constraints on the mobility of robot joints;
c.   Collision avoidance within the robot workspace;
(where $x_i^k$ and $x_i^f$ are the Cartesian coordinates of intermediate and final configurations $C^k$ and $C^k$ respectively).
The process for calculating the whole trajectory between initial and final configurations ( $C^1$ and $C^f$) is based on a second and different optimization problem which can be stated as:

$$\text{Find} \quad q(t), \tau(t), t_f \text{ between each two configurations} \tag{18}$$

$$\text{Minimizing} \min_{\tau \in \Omega} J = \int_{0}^{t_f} 1 \cdot dt \tag{19}$$

Subject to the robot dynamics

$$M(q(t))\ddot{q}(t) + C(q(t),\dot{q}(t))\dot{q}(t) + g(q(t)) = \tau(t) \tag{20}$$

Unknown boundary conditions for intermediate configurations a priori

$$q(t_{int-1}) = q_{int-1} \; ; \; q(t_{int}) = q_{int}$$
$$\dot{q}(t_{int-1}) = \dot{q}_{int-1} \; ; \; \dot{q}(t_{int}) = \dot{q}_{int-1} \tag{21}$$
$$\ddot{q}(t_{int-1}) = \ddot{q}_{int-1} \; ; \; \ddot{q}(t_{int}) = \ddot{q}_{int-1}$$

Boundary conditions for initial and final configuration (used to solve the first and final step)

$$q(0) - q_o \; ; \; q(t_f) = q_f$$
$$\dot{q}(0) = 0 \; ; \; \dot{q}(t_f) = 0 \tag{22}$$

Actuator torque rate limits

$$\tau_{min} \leq \tau(t) \leq \tau_{max} \tag{23}$$

Collision avoidance within the robot workspace

$$d_{ij} \geq r_j + w_j \tag{24}$$

where $d_{ij}$ is the distance from any obstacle pattern $j$ (sphere, cylinder or prism) to link $i$; $r_j$ is the characteristic radius of the obstacle pattern and $w_j$ is the radius of the smallest cylinder that contains the link $i$.

As well, $q(t) \in R^n$ is the vector of joint positions ($n$ being the number of degrees of freedom of the robot), $\tau(t) \in R^n$ is the vector of actuator torques, $M(q(t)) \in R^{n \times n}$ is the inertia matrix of the robot, $C(q(t),\dot{q}(t)) \in R^{n \times n}$ is a third-order tensor representing the coefficients of centrifugal and Coriolis forces and $G(q(t)) \in R^n$ is the vector of gravity terms, and $\Omega$ is the space state in which the vector of actuator torques is feasible. Each time a new adjacent configuration $C^k$ is generated, an uncertainty to be overcome lies in the fact that at this stage we do not know its kinematic characteristics (particularly velocity and acceleration), although we know they should be compatible with the dynamic characteristics of the robot.

It should also be noted when calculating the minimum time between two adjacent configurations that each step starts from a configuration with its kinematic properties known, obtaining the time and the kinematic properties at the end configuration, so that if the dynamic capabilities of the actuators had been exhausted ($\tau_i(t_{int}) \cong \tau_{min}$ or $\tau_i(t_{int}) \cong \tau_{max}$) due to the kinematic properties generated at the end configuration ($q(t_{int})$, $\dot{q}(t_{int})$ and $\ddot{q}(t_{int})$), it would have been impossible to observe constraints on the next generation step of the trajectory $\tau_{min} \leq \tau(t_{int+1}) \leq \tau_{max}$. Finally, by connecting adjacent configurations, the whole trajectory is generated.

The process explained is applied repeatedly to generate adjacent configurations until reaching the final configuration. Finally, by connecting adjacent configurations, the whole trajectory is generated.

## 9.2 Interpolation function

It must be noticed that three types of trajectory spans should be distinguished because of their different boundary conditions: the initial (which contains the initial configuration $C^1$), the final (which contains the final configuration $C^f$), and the intermediate (which does not contain either the initial or the final configuration).

Each pair of adjacent configurations is interpolated using harmonic functions in order to limit the kinematic characteristics of goal configuration so that progression to the following step should be admissible without breaking the dynamic properties. In that way, it is not necessary to previously impose kinematic constraints onto the process. Now, starting from the initial configuration, the harmonic function leads to the knowledge of the kinematic characteristics of the configurations adjacent to it, and so on. And therefore the process of obtaining adjacent configurations can continue until reaching the end. It is true that the results are influenced by the use of different interpolation functions between adjacent configurations. We use harmonic functions because they are capable of limiting the maximum values of velocity and acceleration required for the actuators. So, values for velocities and accelerations are limited. This important trait is deduced from the properties of Fourier series because of the harmonic functions used as interpolation functions and can be expressed by means of their Fourier series, which can ultimately be expressed as

$$f(t) = C_0 + C_1 \cos(t + \theta_1) \tag{25}$$

whose $C_1$ coefficient is the value of the amplitude for the fundamental component and $\theta_1$ is the phase angle. It can be demonstrated that the values of the function are limited to the interval $[C_1, -C_1]$ (the coefficients of the cosine terms). Analyzing the harmonic function on the basis of the type of trajectory span we distinguish three types. We use different interpolation functions to determine their impact on the characteristics of the solution generated. The cases analyzed and interpolation functions for each case are as follows.

a.   Initial span:   Cases A, B and C

In all three cases we have used the same interpolation function for the first span, therefore the procedure to calculate the constants is identical

$$q_{i1} = a_{i1} \cdot \sin(t) - b_{i1} \cdot \cos(t) + c_{i1} \tag{26}$$

with $i = 1..N_{dof}$ and 1 is for the first span. $N_{dof}$ is the number of the robot's degrees of freedom. For this type of interpolation function, velocity and acceleration values are limited by the coefficients $a_{ij}$, $b_{ij}$. The known boundary conditions are three: the initial and final configuration of the interval and the initial velocity. They allow the set of coefficients $a_{ij}$, $b_{ij}$ and $c_{ij}$ to be obtained, which are dependent on time.

b.   Intermediate span.

Three different interpolation functions corresponding to cases A, B and C have been used. To calculate the constants in each case we have proceeded as follows:

b1)  Case A

The interpolation function is

$$q_{ij} = a_{ij} \cdot sin(t) - b_{ij} \cdot cos(2 \cdot t) + c_{ij} \cdot sin(3 \cdot t) - d_{ij} \cdot cos(4 \cdot t) \tag{27}$$

with $i=1..N_{dof}$ and $j=1..N_{span}$. . $N_{span}$ is the number of the span that is being analyzed.
From experience in the resolution of a great number of cases, a polynomial term has been added to ensure the boundary conditions of velocity and acceleration along the trajectory in this span. Velocity and acceleration equations are

$$\dot{q}_{ij} = a_{ij} \cdot cos(t) + 2 \cdot b_{ij} \cdot sin(2 \cdot t) + 3 \cdot c_{ij} \cdot cos(3 \cdot t) + 4 \cdot d_{ij} \cdot sin(4 \cdot t) \tag{28}$$

$$\ddot{q}_{ij} = -a_{ij} \cdot sin(t) + 4 \cdot b_{ij} \cdot cos(2 \cdot t) - 9 \cdot c_{ij} \cdot sin(3 \cdot t) + 16 \cdot d_{ij} \cdot cos(4 \cdot t) \tag{29}$$

Their values are limited by the coefficients $a_{ij}$, $b_{ij}$, $c_{ij}$ and $d_{ij}$. The known boundary conditions are four: the initial and final configurations of the span and the velocities and accelerations at the beginning, and they allow the expressions for the constants $a_{ij}$, $b_{ij}$, $c_{ij}$ and $d_{ij}$ to be determined which, as in the previous case, are dependent on time.

b2)  Case B

The interpolation function is

$$q_{ij} = cos(t) \cdot (sin(t) \cdot (a_{ij} \cdot sin(t) + b_{ij}) + c_{ij}) + d_{ij} \tag{30}$$

Velocity and acceleration equations are

$$\dot{q}_{ij} = cos(t) \cdot (cos(t) \cdot (a_{ij} \cdot sin(t) + b_{ij}) + a_{ij} \cdot cos(t) \cdot sin(t)) - sin(t) \cdot (sin(t) \cdot (a_{ij} \cdot sin(t) + b_{ij}) + c_{ij}) \tag{31}$$

$$\ddot{q}_{ij} = cos(t) \cdot (-a_{ij} \cdot sin^2(t) - sin(t) \cdot (a_{ij} \cdot sin(t) + b_{ij}) + 2 \cdot a_{ij} \cdot cos^2(t)) - cos(t) \cdot (sin(t) \cdot (a_{ij} \cdot sin(t) + b_{ij}) + c_{ij}) - 2 \cdot sin(t) \cdot (cos(t) \cdot (a_{ij} \cdot sin(t) + b_{ij}) + a_{ij} \cdot cos(t) \cdot sin(t)) \tag{32}$$

Their values are limited by the new coefficients $a_{ij}$, $b_{ij}$, $c_{ij}$ and $d_{ij}$. The known boundary conditions are also four: the initial and final configurations of the span and the velocities and accelerations at the beginning. Therefore the constants $a_{ij}$, $b_{ij}$, $c_{ij}$ and $d_{ij}$ can be determined which, as in the previous case, are dependent on time.

b3.  Case C

The interpolation function is

$$q_{ij} = sin(t) \cdot (cos(t) \cdot (a_{ij} \cdot sin(t) + b_{ij}) + c_{ij}) + d_{ij} \tag{33}$$

Velocity and acceleration equations are

$$\dot{q}_{ij} = sin(t) (a_{ij} \cdot cos^2(t) - sin(t) \cdot (a_{ij} \cdot sin(t) + b_{ij})) + cos(t) \cdot (cos(t) \cdot (a_{ij} \cdot sin(t) + b_{ij}) + c_{ij}) \tag{34}$$

$$\ddot{q}_{ij} = 2 \cdot cos(t) \cdot (a_{ij} \cdot cos^2(t) - sin(t) \cdot (a_{ij} \cdot sin(t) + b_{ij})) - sin(t) \cdot (cos(t) \cdot (a_{ij} \cdot sin(t) + b_{ij}) + c_{ij}) + sin(t) \cdot (-cos(t) \cdot (a_{ij} \cdot sin(t) + b_{ij}) - 3 \cdot a_{ij} \cdot cos(t) \cdot sin(t)) \tag{35}$$

Their values are limited by the new coefficients $a_{ij}$, $b_{ij}$, $c_{ij}$ and $d_{ij}$. The known boundary conditions are also four: the initial and final configurations of the span and the velocities and accelerations at the beginning. Therefore the constants $a_{ij}$, $b_{ij}$, $c_{ij}$ and $d_{ij}$ can be determined which, as in the previous case, are dependent on time.

c.  Final span: Cases A, B and C

In all three cases we used the same interpolation function for the last span and therefore the procedure to calculate the constants is identical

$$q_{iF} = a_{iF} \cdot sin(t) + b_{iF} \cdot cos(t) + c_{iF} \cdot sin(t)^2 + d_{jF} \cdot t + e_{iF} \cdot t^2 \tag{36}$$

with $i=1..N_{dof}$ and $F$ is for the final trajectory span.

In this type of span a polynomial term has been introduced, in this case of grade 2, which would ensure the continuity of velocity and acceleration. The velocity and acceleration equations are

$$\dot{q}_{iF} = a_{iF} \cdot cos(t) - b_{iF} \cdot sin(t) + 2 \cdot c_{iF} \cdot sin(t) \cdot cos(t) + d_{iF} + 2 \cdot e_{iF} \cdot t \tag{37}$$

$$\ddot{q}_{iF} = -a_{iF} \cdot sin(t) - b_{iF} \cdot cos(t) + 2 \cdot c_{iF} \cdot (cos(t)^2 - sin(t)^2) + 2 \cdot e_{iF} \tag{38}$$

Their values are limited by the coefficients $a_{ij}$, $b_{ij}$, $c_{ij}$, $d_{ij}$ and $e_{ij}$. The known boundary conditions are five: the initial and final configuration in the last span or interval, the velocity and acceleration at the beginning of the interval and the velocity at the end. These boundary conditions enable the coefficients $a_{ij}$, $b_{ij}$, $c_{ij}$, $d_{ij}$ and $e_{ij}$ to be obtained.

Whenever a new adjacent configuration is generated by solving Eq. (4), a new trajectory span will also be created (by solving the second optimization problem Eq. (17)), and the necessary time $t_j$ to perform the span is then obtained. The joint positions are adjusted using the corresponding harmonic interpolation function again. The solution of equations is obtained by iteration using quadratic sequential programming techniques (SQP) through the mathematical commercial software NAG (Numerical Algorithms Group). An each step of the iterative process it is necessary to recalculate the coefficients of the harmonic interpolation functions used, since they are time functions. To facilitate calculations, each span has been discretized using ten subintervals, so that the kinematic and dynamic characteristics are to be calculated at this discrete set of points. The solution of the optimization process provides the minimum time $t_j$ to go from one configuration to its adjacent one and consequently the joint positions $q(t)$ that must be followed between these two configurations, as well as the necessary torques in the actuators $\tau(t)$ and the corresponding kinematic characteristics $\dot{q}(t)$ and $\ddot{q}(t)$ .

### 9.3 Impact of interpolation function

As it was said earlier, the impact of the interpolation function is very important from the point of view of the robot´s performance. Three types of interpolation functions have been used (A, B and C) for the computation of intermediate configurations (harmonic functions) when using the "simultaneous algorithm. Pure polynomial interpolation functions have been excluded because they exceeded the dynamic capabilities of the actuators and therefore the algorithm failed to reach any solution. Therefore, after having analysed all kinds of interpolation functions, we state that the best of all them C (notice that each actuator has been characterized by the maximum and minimum torque it can provide, see Eq. (23). Nonetheless, both the computational and execution time are very high compared with the results obtained using the "sequential algorithm".

## 9.4 Cost function
An important point of the algorithm is to understand the process by which the algorithm is gradually creating the trajectory. The algorithm works in a discretised workspace (see Rubio et al.,2009) , looking for a trajectory that joins the initial and final configurations by starting from the initial configuration and, on the basis of generating adjacent configurations and branching out from the most promising one, obtaining new configurations until reaching the final one. Therefore, the trajectory contains a discrete set of intermediate configurations. To ensure that the process moves from one configuration to another, that is, that the algorithm branches out from a general intermediate configuration to generate more new adjacent configurations, the uniform cost function is used. The discrete configuration space is analysed as a graph, where the configurations generated are the nodes and the arc between nodes (arc $(i, j)= time(i, j)$ ) is calculated as the time necessary to perform the motion between adjacent configurations. It is desirable that the number of configurations generated is not high and, in addition, that these configurations enable efficient trajectories to be obtained. The process followed to achieve the growth of the configuration space in the search for the final configuration is as follows:
Let $CC = \{C^1, C^2, ..., C^k\}$ be the set of existing configurations at a given instant, and $CR$ the subgroup of $CC$ that contains $r$ ($r < k$) configurations that have still not been used to branch out. Now, it is necessary to follow what is called a branching strategy or searching strategy to select a $C^p$ configuration pertaining to $CR$, from which the algorithm tries to generate another six new adjacent configurations $C^{p+1}$, $C^{p+2}$, $C^{p+3}$, $C^{p+4}$, $C^{p+5}$ and $C^{p+6}$ (according to the technique explained in Valero (2006)), which are new configurations belonging to $CR$, while $C^p$ is taken out of this subgroup. The process finishes when the final configuration is reached. The cost function $c$ $(p)$ used to select a new configuration to branch out from is defined as follows
• Uniform Cost: the time function $c$ $(j)$ associated with the configuration $C^j$ is defined as the minimum sum of arcs that permit the node $j$ to be reached from the initial node

$$c(j) = time(1, j) \qquad (39)$$

And the new branching is started from configuration C p , which meets

$$c(p) = min\left[c(j)\right], \forall j \in CR \qquad (40)$$

When a set of adjacent configurations has been created, Eq. (40) is used to select that one from which the process is expected to branch out again. Given two adjacent configurations the minimum time between them is calculated as explained in Section 4. Time is used to select the new configuration as just explained in Section 5, which is used to repeat the branching process and this in turn is repeated until the final configuration is reached.

## 9.5 Obtaining the trajectory
When the final configuration is reached we know not only the robot configuration through the joint positions $q(t)$ but also the necessary torques $\tau(t)$ and the kinematic

characteristics of the motion $\dot{q}(t)$ and $\ddot{q}(t)$ . The trajectory obtained is of minimum time on the graph generated. To obtain the global minimum time, the process should be repeated with different discretization sizes. The global minimum time is the smallest of all times calculated.

### 9.6 Application and examples solved
This algorithm has been applied to the PUMA 560 robot type, and a great number of examples have been analysed. Four important operational parameters have been monitored: the computational time used in generating a solution, the execution time, the distance travelled (which corresponds to the sum of the whole distance travelled by each significant point throughout the path to go from the initial to the final configuration, measured in meters) and the number of configurations generated. Though the examples, the behaviour of those four operational parameters mentioned earlier when the simultaneous algorithm and the different interpolation functions have been use can be analized. The results obtained show that the worst computational time is achieved when using the interpolation function of case A. Case B and C yield similar results. Also, the results show that the smallest execution time is achieved when using the interpolation function of case C. The smallest distance travelled is achieved when using the interpolation function of case C as well as the smallest number of configurations generated are achieved when using the interpolation function of case C.

## 10. Conclusion

In this paper, two algorithms that solve the trajectory planning problem for industrial robots in an environment with obstacles have been introduced and summarized. They have been called "sequential" and "simultaneous" algorithm respectively. Both are off-line algorithms. The first one is based on an indirect methodology because it solves the trajectory planning in two sequential steps (first a path is generated and once the path is known, a trajectory is adjusted to it). Polynomial interpolation functions have been in this algorithm because they yield the best results. Besides, the trajectories calculated meet constraints on torque, power, jerk and energy consumed. The second algorithm is a direct method, which solves the equations in the state space of the robot. Unlike other direct methods, it does not use previously defined paths, which enables working with mobile obstacles although the obstacles used in this chapter are statics. Three types of interpolation functions have been used for the computation of intermediate configurations (harmonic functions). Polynomial interpolation functions have been excluded from this algorithm because during the resolution phase of the examples, because converge problems in the optimization problem have come up.

The main conclusions are summarized as follows:
a.  The algorithms solve the trajectory planning problem for industrial robots in environments with obstacles therefore avoiding collisions.
b.  It can be applied to any industrial robot.
c.  "Sequential" algorithm:
    c.1. Constraints on the energy consumed must be compatible with the robot´s demanded potential energy, as energy recovery is not considered, as the algorithm works on the assumption that the energy can be dissipated but not recovered.

 c.2. To obtain competitive results in the balance between time cycle and energy consumed, the actuators should work with the maximum admissible value of the jerk so that the robot can work with the desired accuracy.

 c.3. The cubic interpolation function gives the best computational and execution time.

d. "Simultaneous" algorithm: as for the peculiarities of the interpolation functions in relation to the four monitored operating parameters (computational time, execution time, distance travelled and number of configurations generated), the main point is that the best results are obtained when using the interpolation function of case C (taking into account that each actuator has been characterized by the maximum and minimum torque it can provide). With this algorithm the cubic interpolation function does not work because during the resolution phase of the examples, they exceeded the dynamic capabilities of the actuators and therefore the algorithm failed to reach any solution.

## 11. Acknowledgment

This paper has been possible thanks to the funding of Science and Innovation Ministry of the Spain Government by means of the Researching and Technologic Development Project DPI2010-20814-C02-01 (IDEMOV).

## 12. References

Abdel-Malek, K., Mi, Z., Yang, J.Z. & Nebel, K. (2006), Optimization-based trajectory planning of the human upper body, *Robotica*, Vol. 24, n° 6, pp. (683-696).

Bobrow, J.E., Dubowsky, S. & Gibson, J.S. (1985), Time-Optimal Control of Robotic Manipulators Along Specied Paths, *International Journal of Robotics Research*, Vol. 4, n° 3, pp. (3-17).

Chen, Y. & Desrochers, A.A. (1989), Structure of minimum time control law for robotic manipulators with constrained paths, *IEEE Int Conf Robot Automat*, ISBN: 0-8186-1938-4, pp. (971-976), Scottsdale, USA, 1989.

Chettibi, T., Lehtihet, H.E., Haddad, M. & Hanchi, S. (2002), Optimal pose trajectory planning for robot manipulators. *Mechanism and Machine Theory*, vol 37, n° 10, pp. (1063-1086).

Chettibi, T., Lehtihet, H.E., Haddad, M. & Hanchi, S. (2004), Minimum cost trajectory planning for industrial robots. European Journal of Mechanics a-solids 23 (4): 703-715.

Cho, B. H., Choi, B. S. & Lee, J. M. (2006), Time-optimal trajectory planning for a robot system under torque and impulse constraints, *International Journal of Control, Automation, and Systems*, Vol. 4, n° 1, pp. (10-16).

Constantinescu, D. & Croft, E.A. (2000), Smooth and time-optimal trajectory planning for industrial manipulators along specified paths, *Journal of Robotic Systems*, Vol. 17, no 5, pp. (233-249).

du Plessis, L. J. & Snyman, J. A. (2003 ), Trajectory-planning through interpolation by overlapping cubic arcs and cubic splines. *International Journal for Numerical Methods in Engineering*, Vol. 57, n° 11, pp. (1615-1641).

Field, G. & Stepanenko, Y. (1996), Iterative dynamic programming: an approach to minimum energy trajectory planning for robotic manipulators, *Proc. of the IEEE*

*International Conference on Robotics and Automation,* ISBN: 0-7803-2988-0, pp. (2755-2760), Minneapolis, USA, 1996.

Garg, D. & Ruengcharungpong, C. (1992), Force balance and energy optimization in cooperating manipulators, *Proceedings of the 23rd Annual Pittsburgh Modeling and Simulation Conference,* pp. (2017-2024), Pittsburgh, USA, 1992.

Gasparetto, A. & Zanotto, V. (2007),A new method for smooth trajectory planning of robot manipulators, *Mechanism and Machine Theory,* Vol. 42, n° 4, pp. (455-471).

Gasparetto, A. & Zanotto, V. (2010), Optimal trajectory planning for industrial robots, *Advances in Engineering Software,* vol. 41, No 4, pp. 548-556.

Hirakawa, A. & Kawamura, A. (1996), Proposal of trajectory generation for redundant manipulators using variational approach applied to minimization of consumed electrical energy, *Proceedings of the Fourth International Workshop on Advanced Motion Control,* ISBN: 0-7803-3219-9, pp. (687-692), Mie, Japan, 1996.

Kyriakopoulos, K.J. & Saridis, G.N. (1988), Minimum jerk path generation, *in IEEE international conference on robotics and automation,* ISBN: 0-8186-0852-8, pp. (364-369), Philadelphia, USA, 1988.

Macfarlane, S. & Croft, E. A. (2003), Jerk-bounded manipulator trajectory planning: Design for real-time applications. *IEEE Transactions on Robotics and Automation,* Vol. 19, n° 1, pp. (42-52).

Provenzano, S. E. (2001), Aplicación de las ecuaciones de Gibbs-Appell a la dinámica de robots. Doctoral thesis, Universidad Politécnica de Valencia, Spain, 2001.

Rubio, F.J., Valero, F.J., Suñer, J.L. & Mata, V. (2009), Direct step-by-step method for industrial robot path planning, *Industrial Robot: An International Journal,* Vol. 36, n° 6, pp. (594-607).

Rubio, F.J., Valero, F.J., Suñer, J.L. and & Mata, V. (2007), Técnicas globales para la planificación de caminos de robots industriales, *VIII Congreso Iberoamericano de Ingeniería Mecánica in Cusco,* ISBN: 978-9972-2885-31, Cuzco (Peru), Octubre, 2007

Saramago, S.F.P. & Steffen, V. Jr. (2000), Optimal trajectory planning of robot manipulators in the presence of moving obstacles, *Mechanism and Machine Theory,* Vol. 35, pp. (1079-1094).

Saramago, S. F. & Steffen Jr,V. (2001), Trajectory modelling of robot manipulators in the presence of obstacles. Journal of optimization theory and applications. 110(1), 17-34.

Shin, K.G. & McKay, N.D. (1985), Minimum-time control of robotic manipulators with geometric path constraints, *IEEE Transactions on Automatic Control,* ISSN: 0018-9286, pp. (531-541).

Suñer, J.L., Valero, F.J., Ródenas, J.J., & Besa, A. (2007), Comparación entre procedimientos de solución de la interpolación por funciones splines para la planificación de trayectorias de robots industriales, *VIII Congreso Iberoamericano de Ingeniería Mecánica in Cusco,* ISBN: 978-9972-2885-31, Cuzco (Peru), Octubre, 2007

Valero,F. J., Mata V. & Besa A. (2006), Trajectory planning in workspaces with obstacles taking into account the dynamic robot behaviour. *Mechanism and Machine Theory.* Vol. 41, pp. (525-536).

# Permissions

The contributors of this book come from diverse backgrounds, making this book a truly international effort. This book will bring forth new frontiers with its revolutionizing research information and detailed analysis of the nascent developments around the world.

We would like to thank Dr. Ashish Dutta, for lending his expertise to make the book truly unique. He has played a crucial role in the development of this book. Without his invaluable contribution this book wouldn't have been possible. He has made vital efforts to compile up to date information on the varied aspects of this subject to make this book a valuable addition to the collection of many professionals and students.

This book was conceptualized with the vision of imparting up-to-date information and advanced data in this field. To ensure the same, a matchless editorial board was set up. Every individual on the board went through rigorous rounds of assessment to prove their worth. After which they invested a large part of their time researching and compiling the most relevant data for our readers. Conferences and sessions were held from time to time between the editorial board and the contributing authors to present the data in the most comprehensible form. The editorial team has worked tirelessly to provide valuable and valid information to help people across the globe.

Every chapter published in this book has been scrutinized by our experts. Their significance has been extensively debated. The topics covered herein carry significant findings which will fuel the growth of the discipline. They may even be implemented as practical applications or may be referred to as a beginning point for another development. Chapters in this book were first published by InTech; hereby published with permission under the Creative Commons Attribution License or equivalent.

The editorial board has been involved in producing this book since its inception. They have spent rigorous hours researching and exploring the diverse topics which have resulted in the successful publishing of this book. They have passed on their knowledge of decades through this book. To expedite this challenging task, the publisher supported the team at every step. A small team of assistant editors was also appointed to further simplify the editing procedure and attain best results for the readers.

Our editorial team has been hand-picked from every corner of the world. Their multi-ethnicity adds dynamic inputs to the discussions which result in innovative outcomes. These outcomes are then further discussed with the researchers and contributors who give their valuable feedback and opinion regarding the same. The feedback is then collaborated with the researches and they are edited in a comprehensive manner to aid the understanding of the subject.

Apart from the editorial board, the designing team has also invested a significant amount of their time in understanding the subject and creating the most relevant covers. They scrutinized every image to scout for the most suitable representation of the subject and create an appropriate cover for the book.

The publishing team has been involved in this book since its early stages. They were actively engaged in every process, be it collecting the data, connecting with the contributors or procuring relevant information. The team has been an ardent support to the editorial, designing and production team. Their endless efforts to recruit the best for this project, has resulted in the accomplishment of this book. They are a veteran in the field of academics and their pool of knowledge is as vast as their experience in printing. Their expertise and guidance has proved useful at every step. Their uncompromising quality standards have made this book an exceptional effort. Their encouragement from time to time has been an inspiration for everyone.

The publisher and the editorial board hope that this book will prove to be a valuable piece of knowledge for researchers, students, practitioners and scholars across the globe.

# List of Contributors

Juan José González España, Jovani Alberto Jiménez Builes and Jaime Alberto Guzmán Luna
Research group in Artificial Intelligence for Education, Universidad Nacional de Colombia, Colombia

Ebrahim Mattar
Intelligent Control & Robotics, Department of Electrical and Electronics Engineering, University of Bahrain, Kingdom of Bahrain

Jagdish Lal Raheja, Radhey Shyam, G. Arun Rajsekhar and P. Bhanu Prasad
Digital Systems Group, Central Electronics Engineering Research Institute (CEERI)/Council of Scientific & Industrial Research (CSIR), Pilani, Rajasthan, India

Christian Schlegel, Andreas Steck and Alex Lotz
Computer Science Department, University of Applied Sciences Ulm, Germany

Oliver Prenzel and Christian Martens
Rheinmetall Defence Electronics, Germany

Uwe Lange, Henning Kampe and Axel Gräser
University of Bremen, Germany

Matthias Burwinkel and Bernd Scholz-Reiter
BIBA Bremen, University of Bremen, Germany

Claudio Urrea
Departamento de Ingeniería Eléctrica, DIE, Universidad de Santiago de Chile, USACH, Santiago, Chile

John Kern
Departamento de Ingeniería Eléctrica, DIE, Universidad de Santiago de Chile, USACH, Santiago, Chile
Escuela de Ingeniería Electrónica y Computación, Universidad Iberoamericana de Ciencias y Tecnología, UNICIT, Santiago, Chile

Holman Ortiz
Escuela de Ingeniería Electrónica y Computación, Universidad Iberoamericana de Ciencias y Tecnología, UNICIT, Santiago, Chile

A. C. Domínguez-Brito, J. Cabrera-Gámez, J. D. Hernández-Sosa, J. Isern-González and E. Fernández-Perdomo
Instituto Universitario SIANI & the Departamento de Informática y Sistemas, Universidad de Las Palmas de Gran Canaria, Spain

**Eva Volná, Michal Janošek, Václav Kocian and Martin Kotyrba**
University of Ostrava, Czech Republic

**Zuzana Oplatková**
Tomas Bata University in Zlín, Czech Republic

**Francisco J. Rubio, Francisco J. Valero, Antonio J. Besa and Ana M. Pedrosa**
Centro de Investigación de Tecnología de Vehículos, Universitat Politècnica de València, Spain

Printed in the USA
CPSIA information can be obtained
at www.ICGtesting.com
JSHW011811301024
72690JS00002B/47